Checklists and Compliance

The McGraw-Hill *CONTROLLING PILOT ERROR* Series

Weather
Terry T. Lankford

Communications
Paul E. Illman

Automation
Vladimir Risukhin

Controlled Flight into Terrain (CFIT/CFTT)
Daryl R. Smith

Training and Instruction
David A. Frazier

Checklists and Compliance
Thomas P. Turner

Maintenance and Mechanics
Larry Reithmaier

Situational Awareness
Paul A. Craig

Fatigue
James C. Miller

Culture, Environment, and CRM
Tony Kern

CONTROLLING PILOT ERROR

Checklists and Compliance

Thomas P. Turner

McGraw-Hill

New York   Chicago   San Francisco   Lisbon   London   Madrid
Mexico City   Milan   New Delhi   San Juan   Seoul
Singapore   Sydney   Toronto

Cataloging-in-Publication Data is on file with the Library of Congress

McGraw-Hill

A Division of The McGraw·Hill Companies

Copyright © 2001 by The McGraw-Hill Companies, Inc. All rights reserved. Printed in the United States of America. Except as permitted under the United States Copyright Act of 1976, no part of this publication may be reproduced or distributed in any form or by any means, or stored in a data base or retrieval system, without the prior written permission of the publisher.

1 2 3 4 5 6 7 8 9 0 DOC/DOC 0 7 6 5 4 3 2 1

ISBN 0-07-137254-7

The sponsoring editor for this book was Shelley Ingram Carr, the editing supervisor was Stephen M. Smith, and the production supervisor was Pamela A. Pelton. It was set in Garamond following the TAB3A design by Victoria Khavkina of McGraw-Hill's Hightstown, N.J., Professional Book Group composition unit.

Printed and bound by R. R. Donnelley & Sons Company.

McGraw-Hill books are available at special quantity discounts to use as premiums and sales promotions, or for use in corporate training programs. For more information, please write to the Director of Special Sales, Professional Publishing, McGraw-Hill, Two Penn Plaza, New York, NY 10121-2298. Or contact your local bookstore.

 This book is printed on recycled, acid-free paper containing a minimum of 50% recycled, de-inked fiber.

Contents

Series Introduction

The Human Condition

The Roman philosopher Cicero may have been the first to record the much-quoted phrase "to err is human." Since that time, for nearly 2000 years, the malady of human error has played out in triumph and tragedy. It has been the subject of countless doctoral dissertations, books, and, more recently, television documentaries such as "History's Greatest Military Blunders." Aviation is not exempt from this scrutiny, as evidenced by the excellent Learning Channel documentary "Blame the Pilot" or the NOVA special "Why Planes Crash," featuring John Nance. Indeed, error is so prevalent throughout history that our flaws have become associated with our very being, hence the phrase *the human condition*.

The Purpose of This Series

Simply stated, the purpose of the Controlling Pilot Error series is to address the so-called human condition, improve performance in aviation, and, in so doing, save a few lives. It is not our intent to rehash the work of over a millennia of expert and amateur opinions but rather to *apply* some of the more important and insightful theoretical perspectives to the life and death arena of manned flight. To the best of my knowledge, no effort of this magnitude has ever been attempted in aviation, or anywhere else for that matter. What follows is an extraordinary combination of why, what, and how to avoid and control error in aviation.

Because most pilots are practical people at heart—many of whom like to spin a yarn over a cold lager—we will apply this wisdom to the daily flight environment, using a case study approach. The vast majority of the case studies you will read are taken directly from aviators who have made mistakes (or have been victimized by the mistakes of others) and survived to tell about it. Further to their credit, they have reported these events via the anonymous Aviation Safety Reporting System (ASRS), an outstanding program that provides a wealth of extremely useful and *usable* data to those who seek to make the skies a safer place.

A Brief Word about the ASRS

The ASRS was established in 1975 under a Memorandum of Agreement between the Federal Aviation Administration (FAA) and the National Aeronautics and Space Administration (NASA). According to the official ASRS web site, *http://asrs.arc.nasa.gov*

> The ASRS collects, analyzes, and responds to voluntarily submitted aviation safety incident reports in order to lessen the likelihood of aviation accidents. ASRS data are used to:
>
> - Identify deficiencies and discrepancies in the National Aviation System (NAS) so that these can be remedied by appropriate authorities.

- Support policy formulation and planning for, and improvements to, the NAS.
- Strengthen the foundation of aviation human factors safety research. This is particularly important since it is generally conceded *that over two-thirds of all aviation accidents and incidents have their roots in human performance errors* (emphasis added).

Certain types of analyses have already been done to the ASRS data to produce "data sets," or prepackaged groups of reports that have been screened "for the relevance to the topic description" (ASRS web site). These data sets serve as the foundation of our Controlling Pilot Error project. The data come *from* practitioners and are *for* practitioners.

The Great Debate

The title for this series was selected after much discussion and considerable debate. This is because many aviation professionals disagree about what should be done about the problem of pilot error. The debate is basically three sided. On one side are those who say we should seek any and all available means to *eliminate* human error from the cockpit. This effort takes on two forms. The first approach, backed by considerable capitalistic enthusiasm, is to automate human error out of the system. Literally billions of dollars are spent on so-called human-aiding technologies, high-tech systems such as the Ground Proximity Warning System (GPWS) and the Traffic Alert and Collision Avoidance System (TCAS). Although these systems have undoubtedly made the skies safer, some argue that they have made the pilot more complacent and dependent on the automation, creating an entirely new set of pilot errors. Already the automation enthusiasts are seeking robotic answers for this new challenge. Not surprisingly, many pilot trainers see the problem from a slightly different angle.

Another branch on the "eliminate error" side of the debate argues for higher training and education standards, more accountability, and better screening. This group (of which I count myself a member) argues that some industries (but not yet ours) simply don't make serious errors, or at least the errors are so infrequent that they are statistically nonexistent. This group asks, "How many errors should we allow those who handle nuclear weapons or highly dangerous viruses like Ebola or anthrax?" The group cites research on high-reliability organizations (HROs) and believes that aviation needs to be molded into the HRO mentality. (For more on high-reliability organizations, see *Culture, Environment, and CRM* in this series.) As you might expect, many status quo aviators don't warm quickly to these ideas for more education, training, and accountability—and point to their excellent safety records to say such efforts are not needed. They recommend a different approach, one where no one is really at fault.

On the far opposite side of the debate lie those who argue for "blameless cultures" and "error-tolerant systems." This group agrees with Cicero that "to err is human" and advocates "error-management," a concept that prepares pilots to recognize and "trap" error before it can build upon itself into a mishap chain of events. The group feels that training should be focused on primarily error mitigation rather than (or, in some cases, in addition to) error prevention.

Falling somewhere between these two extremes are two less-radical but still opposing ideas. The first approach is designed to prevent a recurring error. It goes something like this: "Pilot X did this or that and it led to a mishap, so don't do what Pilot X did." Regulators are particularly fond of this approach, and they attempt to regulate the last mishap out of future existence. These so-called rules written in blood provide the traditionalist with plenty of training materials and even come with ready-made case studies—the mishap that precipitated the rule.

Opponents to this "last mishap" philosophy argue for a more positive approach, one where we educate and train *toward* a complete set of known and valid competencies (positive behaviors) instead of seeking to eliminate negative behaviors. This group argues that the professional airmanship potential of the vast majority of our aviators is seldom approached—let alone realized. This was the subject of an earlier McGraw-Hill release, *Redefining Airmanship*.[1]

Who's Right? Who's Wrong? Who Cares?

It's not about *who's* right, but rather *what's* right. Taking the philosophy that there is value in all sides of a debate, the Controlling Pilot Error series is the first truly comprehensive approach to pilot error. By taking a unique "before-during-after" approach and using modern-era case studies, 10 authors—each an expert in the subject at hand—methodically attack the problem of pilot error from several angles. First, they focus on error prevention by taking a case study and showing how preemptive education and training, applied to planning and execution, could have avoided the error entirely. Second, the authors apply error management principles to the case study to show how a mistake could have been (or was) mitigated after it was made. Finally, the case study participants are treated to a thorough "debrief," where alternatives are discussed to prevent a reoccurrence of the error. By analyzing the conditions before, during, and after each case study, we hope to combine the best of all areas of the error-prevention debate.

A Word on Authors and Format

Topics and authors for this series were carefully analyzed and hand-picked. As mentioned earlier, the topics were taken from preculled data sets and selected for their relevance by NASA-Ames scientists. The authors were chosen for their interest and expertise in the given topic area. Some are experienced authors and researchers, but, more importantly, *all* are highly experienced in the aviation field about which they are writing. In a word, they are practitioners and have "been there and done that" as it relates to their particular topic.

In many cases, the authors have chosen to expand on the ASRS reports with case studies from a variety of sources, including their own experience. Although Controlling Pilot Error is designed as a comprehensive series, the reader should not expect complete uniformity of format or analytical approach. Each author has brought his own unique style and strengths to bear on the problem at hand. For this reason, each volume in the series can be used as a stand-alone reference or as a part of a complete library of common pilot error materials.

Although there are nearly as many ways to view pilot error as there are to make them, all authors were familiarized with what I personally believe should be the industry standard for the analysis of human error in aviation. The Human Factors Analysis and Classification System (HFACS) builds upon the groundbreaking and seminal work of James Reason to identify and organize human error into distinct and extremely useful subcategories. Scott Shappell and Doug Wiegmann completed the picture of error and error resistance by identifying common fail points in organizations and individuals. The following overview of this outstanding guide[2] to understanding pilot error is adapted from a United States Navy mishap investigation presentation.

> Simply writing off aviation mishaps to "aircrew error" is a simplistic, if not naive, approach to mishap causation. After all, it is well established that mishaps cannot be attributed to a single cause, or in most instances, even a single individual. Rather, accidents are the end result of a myriad of latent and active failures, only the last of which are the unsafe acts of the aircrew.
>
> As described by Reason,[3] active failures are the actions or inactions of operators that are believed to cause the accident. Traditionally referred to as "pilot error," they are the last "unsafe acts" committed by aircrew, often with immediate and tragic consequences. For example, forgetting to lower the landing gear before touch down or hotdogging through a box canyon will yield relatively immediate, and potentially grave, consequences.
>
> In contrast, latent failures are errors committed by individuals within the supervisory chain of command that effect the tragic sequence of events characteristic of an accident. For example, it is not difficult to understand how tasking aviators at the expense of quality crew rest can lead to fatigue and ultimately errors (active failures) in the cockpit. Viewed from this perspective then, the unsafe acts of aircrew are the end result of a long chain of causes whose roots

originate in other parts (often the upper eche-
lons) of the organization. The problem is that
these latent failures may lie dormant or unde-
tected for hours, days, weeks, or longer until
one day they bite the unsuspecting aircrew....

What makes [Reason's] "Swiss Cheese" model
particularly useful in any investigation of pilot
error is that it forces investigators to address
latent failures within the causal sequence of
events as well. For instance, latent failures such

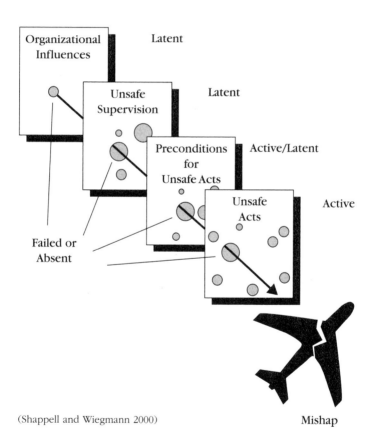

(Shappell and Wiegmann 2000)

as fatigue, complacency, illness, and the loss of situational awareness all effect performance but can be overlooked by investigators with even the best of intentions. These particular latent failures are described within the context of the "Swiss Cheese" model as preconditions for unsafe acts. Likewise, unsafe supervisory practices can promote unsafe conditions within operators and ultimately unsafe acts will occur. Regardless, whenever a mishap does occur, the crew naturally bears a great deal of the responsibility and must be held accountable. However, in many instances, the latent failures at the supervisory level were equally, if not more, responsible for the mishap. In a sense, the crew was set up for failure....

But the "Swiss Cheese" model doesn't stop at the supervisory levels either; the organization itself can impact performance at all levels. For instance, in times of fiscal austerity funding is often cut, and as a result, training and flight time are curtailed. Supervisors are therefore left with tasking "non-proficient" aviators with sometimes-complex missions. Not surprisingly, causal factors such as task saturation and the loss of situational awareness will begin to appear and consequently performance in the cockpit will suffer. As such, causal factors at all levels must be addressed if any mishap investigation and prevention system is going to work.[4]

The HFACS serves as a reference for error interpretation throughout this series, and we gratefully acknowledge the works of Drs. Reason, Shappell, and Wiegmann in this effort.

No Time to Lose

So let us begin a journey together toward greater knowledge, improved awareness, and safer skies. Pick up any volume in this series and begin the process of self-analysis that is required for significant personal or organizational change. The complexity of the aviation environment demands a foundation of solid airmanship and a healthy, positive approach to combating pilot error. We believe this series will help you on this quest.

References

1. Kern, Tony, *Redefining Airmanship,* McGraw-Hill, New York, 1997.

2. Shappell, S. A., and Wiegmann, D. A., *The Human Factors Analysis and Classification System— HFACS,* DOT/FAA/AM-00/7, February 2000.

3. Reason, J. T., *Human Error,* Cambridge University Press, Cambridge, England, 1990.

4. U.S. Navy, *A Human Error Approach to Accident Investigation,* OPNAV 3750.6R, Appendix O, 2000.

Tony Kern

Foreword

Pilots are creatures of habit. It is critically important that we are, because habit patterns allow us to handle the complicated business of operating an aircraft in a *routine* fashion. What looks exceedingly difficult to others we know to be nothing more than a set of patterns practiced until they are embedded firmly in our physical and cognitive memories. This book on checklists and compliance looks at the phenomena of habit patterns, how we can use them effectively, and how and why we deviate. In my humble opinion, it is the most important volume in this 10-book series.

Henry James calls habit the "enormous flywheel of society," pointing out that our reliance and adherence to habit patterns keeps most of us safe, effective, and efficient in running the day-to-day business of our lives. Habit patterns keep us "in the groove" or "on track." As creatures of habit, pilots should be even more aware of this valuable function—but oddly some of us are not and stray from procedure for a variety of unnecessary reasons. When we stray from proven patterns and checklist compliance, we negate this flywheel function and rely entirely on our own wits and judgment. We do so at our own peril and often the peril of others who are flying with us. Simply stated, in the vast majority of cases, deviations from procedure are an unnecessary risk. Yet they continue in large numbers. Why?

Tom Turner takes on the challenge of answering this and other questions about pilot deviations in this groundbreaking book—to my knowledge, the only book on this subject you will find anywhere. As in all good books written by pilots, he begins this one with a war story, a time when something happened to him that shaped the way he would fly for the rest of his life. I, too, would like to share such an event, which occurred a couple of decades ago, an event I was extremely fortunate to survive.

I was a copilot on a KC-135 (a B707 military air refueling tanker) and was doing a takeoff and departure out of Wurtsmith AFB in Michigan. My pilot and I had just returned from an overseas assignment and were extremely proficient—so proficient, in fact, that we had stopped using the normal procedures checklist because we felt we had the whole thing committed to memory. This subtle arrogance nearly killed us.

Immediately after liftoff, the pilot inadvertently began to drain one of our main wing tanks into the aft body tank to burn (he thought he was draining the wing tip reserve tank, a trifling error of about 8000 pounds). Almost immediately, I began to notice two things. First, the aircraft wanted to pitch up, so I applied nose down trim, and, more noticeably, the plane wanted to roll left. I tried to communicate the problem to my pilot, but he dismissed my concern as a weak copilot's flying skill issue. After another minute or so, the roll was becoming increasingly difficult to control, and I was getting scared. I emphatically requested that he take the aircraft to see how bad the problem really was. I still remember the look on the major's face when I handed over the controls. "When did you first notice this?!" he commanded/ screamed/spit. "Right about at flap retraction," I replied, smugly satisfied but still scared.

We immediately seized upon a "split flap" condition as our problem, and, once again without consulting any checklists, began to slow the aircraft toward flap retraction speed. In reality, we were slowing toward minimum in-flight control speed for our worsening condition, and if a 19-year-old boom operator had not intervened with a checklist in hand, two highly skilled, highly proficient, Air Force–trained, steely eyed killers would have planted that baby right into Lake Huron within seconds.

We had stripped ourselves of the protection provided by the checklist and, when Murphy intervened, had blown the call big time. We had created a self-induced emergency. My father used to tell me that God watched out for drunks and fools—this was my first validation.

The true value of this book does not lie in these types of events, however. Tom Turner takes mundane circumstances—the everyday events of pilot and crew— and turns them into golden lessons we can all relate to. He breaks the subject of compliance into critical segments such as fuel-related mishaps, improper takeoff and landing configurations, and the ultimate embarrassment, gear-up landings. He completes the book with a section on suggested techniques and danger signs, to assist all of us in overcoming this subtle temptation to deviate.

Mr. Turner's diverse background has uniquely prepared him for this contribution. His experience includes a stint as the lead instructor of Flight Safety International's Beech Bonanza school, service as a chief pilot for an engine modification firm, and, more recently, being the director of Mastery Flight Training. He has written books and numerous articles. He has done the pilot community an immense favor with this book on checklists and compliance. I am a better pilot today for having read it, and I think you will feel the same when you do so.

Tony Kern

Introduction

Several years ago I flew a turbocharged Beech Bonanza from Wichita, Kansas, to Springfield, Missouri. I was taking a flight student, who was not yet instrument rated and who had left his plane in Springfield ahead of a blinding snow, to his own Bonanza after the weather cleared. The hour-long flight, direct across the Flint Hills and the east Kansas prairie, was uneventful save for a little low-level turbulence and building cumulus clouds near our destination.

We landed. I taxied alongside my student's A36, he got out through the Bonanza's big rear doors, and I picked up my instrument clearance for the flight back to Wichita. I conscientiously worked through the Before Takeoff checklist from the Beech's pilot's operating handbook (POH), then took off for the hour's flight home.

Traffic had picked up and weather was moving in, and I was given several intermediate level-offs and off-route vectors as I climbed out of Springfield. I had to ask for, and was granted, deviations around the larger buildups during my climb to 8000 feet. Finally on altitude and cleared "direct," I settled into a regimen of checking the

Stormscope, scanning into the thick haze and steering around the developing line of storms.

It was only after I'd cruised for 20 minutes or more that I realized I'd forgotten a critical item from the Cruise checklist. My mixture control, always left at "full rich" for takeoff and climb in the turbocharged Bonanza, was still at full rich even this far into the flight. I was cruising along among the clouds, burning about 24 gallons per hour, nearly three-fourths again what the POH said was expected in cruise flight. It was not critical on this particular trip, because I'd been burning from a nearly full wing tank on this leg, and even at 24 gph, I would have made it to Wichita with fuel to spare. But it occurred to me that, had I not noticed my error and if I had been flying much farther than Wichita, I could easily have run a wing tank dry and caused an engine failure, possibly at a critical point in the flight when recovery was difficult.

To that point I was always diligent about using, and teaching the use of, detailed checklists for preflight and pretakeoff activities, and for shutting down and securing the airplane at the end of a trip. I had not until then realized the enormous benefit, however, of *in-flight* checklists as a means of recovering from lapses in memory when weather, traffic, or other distractions interrupted my thought processes and flight procedures. Right then I became a proponent of the use of checklists for *all* phases of flight.

About that same time, I had the honor of speaking before the membership of a large flying club. I talked about the general aviation accident record and pilot decision making, and touched on my growing appreciation of the need for good pilot procedure as a means of overcoming human factors that affect safety.

After my presentation a number of member-pilots came forward with some questions. One gentleman wanted to ask my opinion on something he'd thought about for a long time. "You see," he explained to me, "my parents live on a farm out in western Kansas. My dad used to fly, and he still mows the grass airstrip near the house. It's a little short for the club airplanes we fly, and there's a ditch and a fence on both ends. Instead of landing there, I've been flying to the local airport, but I think I might be able to take off from Dad's strip. The pilot's operating handbook," he went on to say, "doesn't have a short-field takeoff procedure, but I think if I take off with full flaps, get the airplane into ground effect, and immediately retract the landing gear, I'll accelerate quickly enough to climb out of ground effect before I get to the end of the runway. Then I can pull up the flaps and climb out."

I considered his suggestion for a moment and then asked if he'd tried this technique yet. No, he replied, but he'd been thinking about it for a long time. I reminded him about the section of the presentation I'd just given on deviations from standard operating procedures (SOPs) as a common factor in airplane accidents and noted that his "technique" was about as far as could be from the "standard" way he'd been flying the airplane. "It might just work," I quipped, "but I wouldn't want to be the first one to try it."

Pilots' checklists and standard operating procedures are tools pilots can use to counter some of the human factors that contribute to nearly 80 percent of all general aviation accidents. Checklists and SOPs are no substitute for good stick-and-rudder flying or a good instrument scan when the skies go gray. But truly safe pilots approach flying with a level of professionalism that includes checklist use and standard techniques as a means of verifying that nothing will be forgotten. This is especially important when distractions (like the weather and traffic at Springfield) or temptations (such as the grass strip at the farm in Kansas) cause you to change the way you fly your airplane. It may be that a large number of aviation accidents, like fuel exhaustion or failure to clear objects on takeoff, are really more checklist- or procedure-related mishaps.

"Checklists" and "Do Lists"

Think back to your very first flying lesson. Chances are, you weren't at all familiar with flying an airplane, and the cockpit of even a minimally equipped trainer seemed daunting and complicated. You first began to make sense of it all when your certified flight instructor (CFI) produced a checklist, usually a faded, one-page, beatup sheet of laminated paper that listed, in order, all the things you need to do to fly safely.

You went through the checklist together with the CFI. He or she had you read a step, and then either performed the action or, if you had a good teacher, let you do it yourself. Soon the engine was running and the oil was warming, and after a later-to-seem-logical procession of checklist steps, you'd checked the systems, configured the airplane, and headed aloft.

What you were doing with your instructor on that first flying lesson was using a "checklist," or what some people call a "do list." A do list is meant to be run step-by-step, with a pause after each physical action to read the next step, then time for you to do the next action before reading yet another step. This procedure is repeated until everything on the checklist is complete. After some time you likely found that you knew in exactly what order the do list was arranged, and soon afterward you might have felt tempted to fire up, check out, configure, and take off in the airplane without ever looking at the list.

That's the way most of us learned to fly.

Consider this, then: The U.S. military started requiring its pilots to use printed checklists in the late 1930s.

With worldwide hostilities looming, the Army Air Corps began the process of procuring long-range bombers, intended to be used against ships that might threaten America's shores. The Boeing Aircraft Company fielded its Model 299 in the competition for new bombing aircraft, a four-engine behemoth later to be known as the B-17 Flying Fortress. Under intense competition and tremendous public scrutiny, one of the prototype YB-17s crashed on takeoff during a test flight at Ohio's Wright–Patterson Army Airfield. Why did the YB-17 crash, almost leading to the cancellation of what would be, if not the most numerous, at least the most celebrated U.S. bomber of World War II? The pilots, in a hurry to get their ship airborne and likely confident in their ability to pilot what for its day was one of the biggest airplanes flying, forgot to disconnect the tail control gust locks before taking off. The big silver ship was uncontrollable on takeoff, and it crashed into a fiery heap on a low rise at the end of the runway.

If professional military-aircraft test pilots could forget such a critical step in their takeoff preparations correctly, reasoned Boeing and Army crash investigators, then any "line" aviator could too. The best solution was to make a list of all the things a pilot needed to do to prepare for takeoff and train all pilots to use the list to avoid missing anything. From there it was a quick leap to creating checklists for all phases of flight, not just the takeoff, and to requiring pilots to use the checklists.

Even if you never fly anything more complex than a light training airplane, it's vital that you properly set up the airplane for takeoff, cruise, descent, and landing. Even the simplest airplanes now have these sorts of checklists in their pilot's operating handbooks. But if you're like most pilots, with a little familiarity comes the temptation to put aside the checklist, to fly from memory. And so most pilots do not know how to properly use a checklist.

Again, your introduction to checklist use likely was really a lesson in using a do list—running step-by-step through a printed procedure that outlines actions in a process that is unfamiliar to you. And rightly so, after a short while you didn't need to be told how to set up the airplane.

But what happens if, like the crew of that YB-17 so many years ago, outside pressures cause you to forget critical flight items or even less crucial items of convenience or comfort? What will prompt you to review your actions, to catch something that you may have missed?

You can use checklists for what they were intended: a check to make sure you haven't missed anything. Go ahead and start up the engine without the printed list if you want—but once you've verified it's running smoothly and the oil pressure is coming up, glance at the Start checklist to make sure you've not missed anything. Out in the run-up area, advance the power and check magnetos, carburetor heat, propeller action—anything called for in the type of airplane you're flying—and then run through the Run-Up checklist to double-check your actions. Complete your preflight preparations, but don't take the runway before checking off those items you've done on the printed checklist in case there's something you've forgotten.

You can use a combination of printed and mnemonic, or memory, checklists if you like. Mnemonics are especially helpful if there are a few, short items you need to accomplish in a high-workload phase of flight, such as just before takeoff or just before landing, when you really don't have time to reference a printed checklist. For instance, I use the acronym FLATS just before lining up for takeoff in a high-performance airplane. In my case, FLATS stands for

F   Flaps up, cowl Flaps open

L   Lights as required, door and window Latches secure

A   Avionics set, transponder to Altitude reporting

T   Trim set, Time recorded

S   Switches, especially fuel pumps and anti-ice protection, as required

You might come up with something even better for your airplane or your operations.

Probably the most common mnemonic used in retractable-gear airplanes is GUMPS. Used as a Before Landing checklist, GUMPS stands for

G   Gas to fullest main tank, as required by the pilot's operating handbook

U   Undercarriage, or landing gear, down and locked

M   Mixture full rich or as required by field elevation

P   Propeller control fully forward

S   Switches as required for landing

A properly executed mnemonic checklist does wonders to confirm you've accomplished critical flight items when you're otherwise too busy to refer to a checklist.

Take this checklist technique with you throughout the flight. Established in climb and trimmed for hands-off flight, and after confirming there's nothing else that's immediately demanding your attention, run through the Climb checklist or some mnemonic of your own to make sure you've retracted the flaps, turned off the landing light, or whatever else needs doing in the airplane you're flying. After you've transitioned to cruise flight and think you've done everything you need to do, refer to the

Cruise checklist; you may find that, as in my experience in the A36TC off of Springfield, workload and weather caused you to forget something critical. Ready to descend? Get established and trimmed for descent, then look at your Descent and, as appropriate, Approach or Before Landing checklists to make sure you've not missed anything and to prepare yourself for the next phase of flight. Clearing the runway, try a mnemonic (I use the same FLATS, changing the position of controls as correct for ground movement) so you can clean up the airplane while still scanning outside for obstructions. Shutting down? Make your last action in the cockpit a run-through of a Shutdown and Securing Airplane checklist—do this and you won't forget to switch off the battery master switch ever again.

Use checklists the way they were intended—as your standard operating procedure and the only way you can envision flying. You'll quickly be able to notice when stress, fatigue, or workload increases the risk of your flight, and more importantly, you'll have a well-rehearsed means of combating these adverse factors that contribute to so many airplane incidents and accidents.

It's been said that it's not bad piloting procedures that cause so many aviation accidents. Rather, it's that many pilots have no set procedures at all. Using checklists and standard operating procedures will make you a safer and more professional pilot, and may reduce your cockpit workload enough to make flying even more fun.

In the following chapters, we'll look at case studies from the Aviation Safety Reporting System (ASRS), the National Transportation Safety Board (NTSB), and other sources to see how good checklist and standard operating procedures can help you enjoy years of safe flying.

1

Fuel-Related Mishaps

Fuel-related accidents are some of the most flagrant results of improper or interrupted pilot technique. Sometimes a fuel emergency arises when a pilot takes off with too little fuel for the trip. As in the Springfield flight example cited in the Introduction, however, fuel-related mishaps in many cases don't result from a conscious decision to take off with too little fuel to complete a planned flight. Instead, distractions, omissions, or changes in the plan often contribute to fuel starvation or exhaustion.

It's helpful here to define fuel exhaustion and fuel starvation. *Fuel exhaustion* is the "running out of gas" most people envision when they consider a fuel-related accident. In a *fuel starvation* mishap, on the other hand, there is still fuel available in a tank somewhere aboard the aircraft, but for some reason one or more of the engines is not getting fuel.

Factors contributing to fuel starvation are

1. Poor pilot monitoring of fuel load, burn, and fuel tank selection
2. Lack of knowledge of the airplane's fuel system and limitations (such as the common limitation that auxiliary tanks are for straight-and-level flight only)
3. Contamination of one or more, but not all, of an airplane's fuel tanks

Historically, the instances of general aviation fuel mismanagement are split almost evenly between fuel starvation and fuel exhaustion.

Case studies show that it's usually an outside factor that leads to a fuel-related accident. Sometimes the result of these outside factors is spectacular or tragic enough that it appears in National Transportation Safety Board (NTSB) findings or even the popular media. More often, though, these factors give us a "near miss," an incident that is not necessarily reportable but that still holds a valuable lesson for us. In the interest of general safety (and often to evoke the enforcement immunity provided by volunteering such reports), some of these near-miss reports make it into the Aviation Safety Reporting System (ASRS). As is the norm for ASRS records, the vast majority of *reported* near misses come from professional pilots flying airline equipment. The experiences of these pilots, however, are a good lesson for those of us flying light airplanes for pleasure or personal business. If a well-trained airline crew with a large ground support team and a system of numerous checks and safety procedures can have a fuel-related emergency, then it is even more likely that distractions or lapses in procedure might catch up with a single pilot operating more or less on his or her own. With that in mind, let's review a few recent ASRS fuel-related accidents.

Case 1: Commuter Crew Fails to Fuel before Takeoff (ASRS Accession Number 370145)

The crew members of a British Aerospace BA31 twin turboprop arrived on time for preflight preparations and briefing, obtained standard weather information, and received release for the first flight of their shift. Part of this release included a fuel order that needed to be filled before takeoff.

The crew preflighted the airplane and settled into the cockpit. Calling for departure clearance, the copilot noticed that a ground support van at an adjacent gate area was spilling fuel from its filler hole. The copilot reported this on the clearance delivery frequency, and then, since he was told he would have to wait for departure clearance, he and the captain began running the Before Start checklist up to the point where takeoff fuel level was confirmed. Since its airplane had not yet been fueled, the crew paused at this step and allowed the gate agent to board passengers into the cabin.

While the passengers were boarding, the crew members watched a Crash Fire Rescue (CFR) truck pull first into their gate and then over to where the van was spilling fuel. Distracted by the emergency response and (as the reporting pilot commented) "checklist forgotten," the crew started engines, obtained instrument clearance, taxied to the runway, and took off.

"As the gear retracted," later writes the reporting crew member, "I looked at the fuel gauges and realized I hadn't received fuel, and I didn't have enough to [make it to] my destination. The last item of the [forgotten] Before Start checklist," he continues, "is to check fuel quantity," which the crew had not done. The flight returned to the departure airport without incident, re-fueled, and then flew safely to its planned destination.

The crew critiques its performance: "Due to the distraction" of the fuel spill and the Crash Fire Rescue response "and lack of sleep the night before, [the crew] forgot to finish [the Before Start] checklist. This is very easy to do, [and] very bad to do. Usually, if interrupted in my checklist, I start over [from the beginning of the procedure], but this time I did not. Nothing bad happened except a broken ego." The reporting pilot theorizes that the crew's lapse in procedure could have been far worse. "It could have been the flaps [we] forgot," which might have prevented a safe takeoff, he writes on the ASRS form.

"Checklists are there for a reason," the reporter concludes, a reason "which shouldn't be neglected."

Preparation

How might the crew have been better prepared to avoid this emergency? The key is what's sometimes called "checklist discipline"—a rigid adherence to procedures that ensures no items of the checklist are missed. After all, the reason we have checklists is to make certain we don't forget to perform tasks critical to the safe outcome of a flight. Maybe the crew could do something tangible that reminds it a checklist is incomplete, like physically placing the checklist ahead of the power levers or leaving the door between the cockpit and cabin open.

The crew knew before boarding the Jetstream that it needed fuel before leaving. This itself should have triggered a mindset that the crew should be doubly careful about completing checklist items. Something unusual in or near the airplane, like the spilling fuel and Crash Fire Rescue response in this incident, should automatically alert the crew to the need for increased vigilance.

In-flight

The Jetstream's crew acted properly as soon as it detected its mistake. The crew noticed the fuel gauges and detected the problem, obtained a clearance to return to the airport, landed, and refueled. Pilots who notice an abnormality that cannot be immediately corrected from the cockpit should not delay in getting the airplane back on the ground.

Post-flight analysis

The ASRS write-up testifies to the reporting pilot's amazement and remorse that the crew could have missed something so critical before taking off. It's "very easy to do," the pilot writes, and the missed item could just as likely have been something with an even more immediate effect on flying safety.

Certainly this particular pair of pilots learned a valuable lesson, and we're indebted to them for making such a complete report available through the ASRS. Hopefully, airline crews, and pilots everywhere, are aware of how easily distraction can affect flight safety and how good checklist procedures can save us from the perils of diverted attention.

Lessons learned

What about a single pilot flying a light airplane in a similar case? Although conscientious lightplane pilots visually check fuel levels before flight, in some airplane types it's impossible to see fuel in the tanks when the tanks are much less than fully filled. A pilot who plans a flight with partially filled tanks needs to be especially careful to confirm that a fuel order was actually carried out. Stay with the airplane until it's fueled, if you've ordered something less than a "top-off" of the tanks. Double-check the fuel payment receipt to ensure that what you ordered is what was put in the tanks. Be especially careful to check the cockpit fuel gauges before taking off.

Distractions can be very powerful, especially when they include strong visual cues like flashing lights and emergency vehicles. Remember that your job on the ramp is not to watch the response outside, it's to make sure your airplane is properly set up for flight. Use distractions themselves as a reminder to more closely process checklists, so that you can catch any critical flight items that might be missed.

The best lesson from this example is only sometimes related to a fuel emergency. If something distracts you from completing a checklist, even momentarily, start over at the top of that checklist and make certain you've accomplished every item. You don't need to redo anything you're sure you've done, but reaccomplishing the checklist will remove any doubt.

Case 2: Distraction Prevents Fueling before Taxi (ASRS Accession Number 385425)

The jet captain and his copilot completed their walk-around of the aircraft about an hour before departure. They noticed the jet's fuel ticket "hanging from [the] wing fuel panel," standard procedure to indicate that fuel had been ordered to be loaded before the flight. The captain, who completed the ASRS report, sat in the cockpit for a while, then left and returned "several times" to check weather, to obtain his company release for flight, and to use the restroom. On one trip he "noticed [a] fuel truck was parked next to [the right wing]," and he "assumed we were getting fuel" loaded into the aircraft. Only later did the captain learn the fuelers were "waiting to fuel a different aircraft."

Back in the cockpit, the copilot called air traffic control (ATC) for departure instrument clearance and was told that the clearance was "on request." The crew then began the Before Start checklist.

ATC radioed back that there was no flight plan filed "in the system" for the trip. The captain left the cockpit, went into the operations office, and called dispatch to direct refiling of the flight plan. When the captain got back to the jet, the ground crew indicated it was ready for the aircrew to start the right engine. The crew completed the "boxed item checklist," a sort of abbreviated procedure for engine start. The captain later writes in his ASRS report that he "did not realize we had never finished [the] Before Start checklist," which included a fuel quantity check.

"Everything proceeded normally from then until taxi," reports the captain. "After receiving taxi clearance I started to taxi off [the] ramp and realized we had 1600 pounds of fuel [on board], not the 2600 pounds" required for the flight. The captain instructed the flight officer to call ground control and request a return to the gate; the captain then called company operations and told it fuel was needed.

Returning to the gate, the captain "ran into operations, called company dispatch, told them what happened," and obtained an amended release. The aircraft was properly fueled, and the flight then proceeded normally to its destination.

Commenting on the event, the captain writes that the "causes of [his] oversight were getting interrupted during the [Before Start] checklist, not finishing the [Before Start] checklist immediately" when interrupted, and "not doing a final walk-around on the last return to the aircraft," which would have revealed the fuel order ticket still attached to the airplane's wing.

Preparation

This is another case of an interrupted Before Start checklist, where the critical fuel status check was missed. As the reporting pilot wrote to the ASRS, proper adherence to checklist procedures would have prevented what could have progressed into a serious fuel emergency. Only the crew's good check of fuel status after engine start and during the taxi-out kept the flight from departing with a dangerously low amount of fuel.

Proper crew checklist training and procedure, along with a technique for dealing with an incomplete fueling discovered during the walk-around (such as a flag or receipt the crew must carry until it's taken by a fueler who has filled the fuel order) would make this sort of incident much less likely. There's no substitute, though, for simply sticking with a checklist until you're sure it's complete, to confirm the fuel status before starting engines and taxiing away.

In-flight

This trip never got airborne before the low fuel state was discovered, and the crew did apparently use good procedure in identifying the situation and promptly returning to the gate.

Post-flight analysis

The captain who reported this incident made a couple of recommendations, which point to better adherence to checklist procedures and not "thinking [we] had completed a checklist which we had not," to avoid this type of mishap. Either finishing the Before Start immediately before dealing with the distraction or performing a final exterior walk-around before pushing back for departure would likely have revealed the discrepancy before taxi.

Lessons learned

The captain reported that he was in and out of the airplane several times before the first engine start and later, after finding the low fuel state and taxiing back in, that he "ran" into operations for what were likely a couple of heated telephone conversations. If you find yourself going to and from the airplane several times, or if you ever feel under enough time pressure that you "run" to or from the aircraft, be especially careful to review and fully complete all checklists. It's these sorts of pressures and movements that can cause you to forget certain tasks, miss checklist steps, or fail to complete checklists altogether. Since checklists are there primarily to help you recover from the effects of your omissions, you need to be especially vigilant about "missed items" when you're rushed or distracted away from completing checklists. Be especially cautious if you ever find yourself *running* around an airplane!

Case 3: Quick-Turn Missed Fueling (ASRS Accession Number 424986)

The crew of a commuter jet requested a "quick turn," or a rapid turnaround including refueling, to stay on a tight flight schedule. "In an effort to get the flight out on time," writes the reporting pilot in the ASRS, "we rushed through the checklist." During the procedure, the pilot recounts, a cabin crew member interrupted the pilots, and this contributed to the pilots missing the critical fuel-state check before pushing back and taking off. By rushing through the checklist and because of the interruption of the checklist items' flow by the cabin crew, the pilots "missed that the aircraft had not been refueled." They noticed the omission after takeoff and returned to the departure airport for fuel.

"In my best estimation," critiques the reporting pilot, "this error occurred because a checklist was interrupted by [the] cabin crew and an item was missed on the checklist as a result. For myself, I will never [again] allow a checklist to be interrupted without starting it over from the beginning."

Preparation

Knowing that it needed to land, refuel, and depart quickly, the crew of this jet might have made a conscious decision to more closely check the critical flight items before starting engines and pushing back for the trip. We can't tell from the ASRS narrative whether the airline provided the crew with an abbreviated "quick-turn" checklist covering the critical steps in a format that contributed to a rapid departure, instead of adding to the stress of trying to take off on a short schedule using the full-procedure checklist.

The pilots may also have signaled cabin crew members before landing for the quick-turn, or at some point during the short time they spent on the ground, that interruptions would not be permitted, but instead the pilots would prompt the cabin crew for any questions at a pilot-determined point on the checklist.

Meanwhile, good coordination between cockpit crew, dispatchers, and fuelers should have prevented a pushback before adding fuel to the aircraft.

In-flight

Again, a good cross-check by the cockpit crew during flight and a prompt decision to return to the departure airport for the neglected fueling prevented a possible disaster.

Post-flight analysis

The reporting pilot provided a good self-critique in this incident. The most salient point is that a checklist, when interrupted, needs to be started again from the top.

Lessons learned

Remember, checklists are designed to prevent you from forgetting things. If a checklist itself is not completed before it's put aside, then picking it back up and starting in midstream does not provide the cross-check and assurance that all the flight-critical items have been addressed. If a checklist doesn't provide you that assurance, then the checklist is useless.

Although ASRS narratives most commonly come from airline crews, sometimes they are filed by general aviation pilots as well. From general aviation reports we learn that lightplane pilots are not immune to checklist-related fuel problems.

Case 4: Fuel Starvation after the Fly-In (ASRS Accession Number 409069)

The pilot of a classic Bellanca 14-13-3 had attended the famous Experimental Aircraft Association (EAA) Convention at Oshkosh, Wisconsin. He had camped by his airplane at the Appleton airport and, by his own admission, "had slept poorly" over the two nights he attended the event.

"I had flown the evening before" the incident, the pilot notes in his report, "and had been distracted [by heavy traffic in the airport traffic pattern] from changing from my auxiliary fuel tank to a main wing tank prior to landing. I preflighted [the Bellanca] prior to departure and checked fuel in the wing tanks, but could not visually check the auxiliary tank [fuel level]."

Departing as the last in a five-airplane formation takeoff, "I had trouble with my aircraft battery on start and had to...pull the propeller through [by hand] to start the engine. I reached the run-up area and did my run-up without my checklist," when he missed moving the fuel selector to a wing tank, as prescribed by the airplane's flight manual. "I was rushed because the last plane in the flight was waiting for me.

"I usually do a configuration check prior to takeoff— 'Trim,' 'Magnetos,' 'Carburetor Heat,' 'Flaps,' and 'Fuel.' I got to 'Flaps' and stopped [because the airshow] runway director was directing me at that time.

"I took off on the auxiliary tank. I ran out of fuel at 300 feet (above ground level) after I had retracted the gear and flaps. I switched to a wing tank and started pumping the [manual] wobble fuel pump, but the engine didn't pick up. I selected the landing gear 'Down' but they [*sic*] didn't get down prior to [the] landing on the runway. Gear in transit, engine stoppage, curled propeller, no injuries. The FAA inspectors on the scene classified it as an 'incident.'"

The reporting pilot enumerated the lessons he learned: "(1) Never run any fuel tank below ¼ (full). (2) At unusual events, double-check yourself. (3) Take all the time you need, use your checklist, and don't omit your usual procedures."

Preparation

It was an exciting time. The pilot was participating in one of the biggest events in aviation. More so, he was demonstrating his unique airplane as part of the show—flying before the day of the incident, and then departing as part of a five-ship formation. There was a *lot* of opportunity for distraction.

The accident really started the day before the gear-up landing, when the swirl of activity in the pattern around the airport contributed to a violation of the pilot's standard operating procedure (SOP)—switching to the wing tank for landing, which is also a checklist step from the airplane's flight manual. Further stress brought on by coordination of the formation takeoff, compounded by the dead battery, the need to "hand-prop" the Bellanca, and the urgings of the airshow runway director, caused the pilot to feel rushed. He missed critical flight items on his Starting and Before Takeoff checklists and violated his own standard operating procedure (the before takeoff final configuration check) and that of the airplane manufacturer (use fuel from the main wing tanks for takeoff).

In-flight

The pilot making this report had only a few seconds of "in-flight" time to respond to the emergency, and it appears he did about everything he could when the engine quit. To his credit, he kept the airplane under control and landed straight ahead on the remaining runway, so he was able to walk away without any physical injury.

It may be that the pilot normally retracted the Bellanca's landing gear at the same point during takeoff as he did in this instance. However, it also may be that, normally, the pilot would delay gear retraction until there was no usable runway remaining—accommodating the typically slow gear extension time of most early-generation, light retractable-gear airplanes. If that's the case, he may have "snatched up the gear" early in this case as part of the élan of the formation takeoff. He was ready to manually pump the auxiliary fuel ("wobble") pump, which shows good situational awareness and disciplined cockpit procedure. What is likely is that this was an otherwise excellent pilot, very familiar with his airplane and extremely competent as its master, who was nonetheless overwhelmed by the excitement of the airshow and the formation flight.

Post-flight analysis

In the final analysis we can chalk this incident up to distraction and the breakdown of good pilot practices the distraction caused. The pilot summed up his experience in three lessons he learned ("Never run any fuel tank below ¼," "at unusual events, double-check yourself," and "take all the time you need, use your checklist, and don't omit your usual procedures"). These are the foundation of a great set of personal standard operating procedures.

Lessons learned

This is the lesson for all pilots—no matter how expert we are, we all still need the protection of well-practiced procedures and the backup of good checklist use. All pilots need to try to foresee what possible distractions or external stresses might affect a flight and to resolve not to let those stresses dictate the way they will fly airplanes. Airshows and fly-ins almost always provide takeoff and landing distractions and involve operation using procedures foreign to our everyday modes of operation. Only good checklist habits and strict adherence to standard operating procedures can protect pilots from the distraction and stress of these unusual times. Checklist use and other standard operating procedures need to be well thought out and practiced to the point that there is no other way the pilot would conceive of flying the airplane.

Case 5: Multiengine Instruction Runs Dry (ASRS Accession Number 407825)

The reporting pilot was conducting flight instruction in a light piston twin. The purpose of the flight was a mid-course check of progress and proficiency of another instructor's students, a "phase check" that is standard procedure in the more professionally run flight schools. While the pilot was entering a practice holding pattern in visual meteorological conditions, the left engine lost power. "Immediately I took the controls," reports the instructor to the ASRS, "and performed the appropriate responses for loss of engine power." This included troubleshooting for possible causes, and the instructor pilot detected the left engine fuel selector in the "crossfeed" position and the right engine fuel selector "on"—the left engine, as well as the right, was drawing fuel from the right wing tank.

Meanwhile, the instructor found the "left engine fuel gauge reading empty and the right gauge reading full." Because the airplane was below 3000 feet above ground level, within five miles of a large, controlled airport, and the airplane's limitation that crossfeed be used only for level flight, reports the instructor, "I opted to feather the left engine [*sic*] and continue to [the airport] VFR, for a single-engine landing. I declared an emergency and, using my student to operate the radios, we proceeded direct" to a landing "without further incident."

The instructor describes some of the factors that led to this incident: The flight was a "phase check" with another instructor's student, who had habits and patterns with which the instructor was not familiar. "During the aircraft taxi-out and run-up I missed the student putting the fuel selectors back to 'on' during the Before

Takeoff checklist. During my cruise (instrument) scan and backup of the student's operation, I did not see the left fuel gauge depleting due to the proximity of the gauge [to] the left side of the control yoke. Before takeoff and in cruise I did not notice the fuel selector in crossfeed. In the terminal area we had multiple distractions with other aircraft…in the pattern."

Preparation

Human beings are creatures of habit, and the vast majority of aviation accidents result from interruptions in the pilot's habit patterns. The first flight together of two persons, a pilot and instructor, who will function as a team in conducting a safe flight, is an especially risky event. A multiengine student and an inexperienced multiengine instructor raise the potential risk to a high level.

Two pilots flying together for the first time need to take additional time before getting in the airplane to brief each other on the flight's conduct and to make sure they agree on the basic procedures that they'll use in the air. Standardized checklists should be a big part of training in and flying all airplanes, but even more so when a pilot is being introduced to multiengine airplanes. Certainly, part of this mission was to verify the progress and training of the multiengine student, so the instructor needed to "sit quietly and watch" to some extent. But the instructor needs to be alert and active in watching the student and intervene in an instructional manner at the first sign of a deviation from standard operating procedures.

In-flight

Instructional flight in a complex airplane is a very busy activity. The student is likely working near his or her current limits of airplane control and understanding and may fixate on tasks immediately at hand to the exclusion of actively looking for potential future problems. The instructor is busy watching the student, critiquing performance, relating new material or hints for better technique, navigating, advising ATC of the flight's progress and negotiating for routes or approaches, and watching for other air traffic. Troubleshooting future trends and problems fall far down on the list of an instructor's immediate priorities, and if the traffic is heavy, the student especially challenging, or the instructor inexperienced, scan and interpretation of status instruments like fuel gauges are often done hastily or ignored.

Strict adherence to standard operating procedures, including printed checklists to back up memory procedures, is the only way to make such an inherently risky activity safe.

To his credit, the instructor seemed to be familiar with the airplane's fuel-related flight limitations and did not hesitate to declare an emergency when the power loss occurred. Instructor and student got the training twin quickly on the ground, and but for the instructor's ASRS report, we would never have learned the lesson of this flight.

Post-flight analysis

The reporting pilot critiques himself: "The flight instructor should continuously back up the student at all times during flight operations, especially during the Before Takeoff checks. Advanced training instructors in multiengine aircraft should be extremely proficient. My low experience with non-multi-rated pilots contributed to the outcome. The student and instructor both learned very valuable lessons regarding checklist discipline and constant vigilance during all phases of aircraft operation."

He debriefed the student as well: "The student mentioned that his previous training was based on using instrument panel-placarded Before Landing checklists. The aircraft checklist incorporates 'Fuel Selectors On,' whereas the instrument panel placard for the Before Landing checklist does not. I critiqued [to the student] that after landing the [printed] checklist should be used as a back-up" to the placard steps.

Lessons learned

Inexperience. Complacency. High-risk activity. Conflicts between printed and placarded checklists. Lack of agreement on standard operating procedures. We have the opportunity to learn from this incident, and I'm certain the reporting pilot is a better instructor for having been through it.

Case 6: Frozen Fuel (NTSB Report Number NYC00LA070)

A commercial pilot escaped injury when the Cessna 150 he was flying ran out of fuel, necessitating a forced landing shortly after takeoff from an airport in Wisconsin. The Cessna was substantially damaged in the crash.

"According to the pilot," states the NTSB Factual Report, he "performed a preflight inspection per the [aircraft] owner's manual. The fuel quantity was 'half full.' The wing [fuel] tank strainer drains were 'frozen stuck,' and the pilot did not force them open. The pilot drained about six ounces of fuel from the fuel strainer" beneath the engine cowling "and found 'no contaminants.' He also noted that 'the fuel vent next to the pitot tube appeared to be open.'

"The pilot started the airplane, taxied it to the runway, and performed a run-up. After that, he back-taxied down

Runway 25, and made a right-crosswind takeoff. After takeoff, 'climb was interrupted seconds into the flight by the engine dying. The altitude was about 125 feet' above ground level. The pilot 'briefly pumped the throttle with no effect on the windmilling engine.'

"The pilot further stated: 'I then saw [a] big white snowy area at the end of Runway 25...and maneuvered with a right turn away from the runway and then a left turn toward the [snowy] area. I was gliding the plane down. With airspeed now getting low the plane would not quickly recover from the slip and touched left wingtip first with the nosewheel collapsing second. The plane was yawing now to the left; the right wingtip dug into the snow decelerating the plane rapidly and yawing it to the right.'"

Preparation

Although the pilot in this incident was conscientious about performing a preflight inspection, the weather was cold enough to freeze the Cessna's fuel strainer drains shut, and the pilot decided to take off without "sumping" the fuel. He did check the fuel at the gascolator drain, the lowest point of the fuel system, but it takes time for contamination to flow down to the bottom of the fuel lines.

More importantly, in this case, ice crystals, water frozen in the fuel system, may float at the *top* of the fuel tanks and not make it down to the gascolator drain or even the wing-mounted strainers. Aviation fuel itself will not freeze at natural outside temperatures, so frozen fuel sump drains point to at least the possibility of water in the fuel system, which is freezing into crystals. Just as water in the fuel can interrupt power, so can crystals of ice—and frozen drains suggest ice in the fuel. The only way to be sure is to thoroughly warm the airplane in a heated hangar and use the thawed strainer drains to check for water in the tanks.

In-flight

Faced with an engine failure immediately after takeoff, the pilot in this mishap did well to maintain control and pick an appropriate place to land. Environmental conditions (surface snow and a crosswind) made the off-airport landing dangerous, but the pilot's proper controlling (and not just a little luck) kept him from being hurt.

Post-flight analysis

Had the pilot recognized that frozen fuel drains equate to possible ice in the fuel itself and that ice in the fuel can cause engine failure, then he may have delayed flight and avoided this mishap.

Lessons learned

If you're unable to accomplish a step on a checklist because part of the airplane is unable to work as designed, that's a strong argument for not flying the airplane at all until the discrepancy can be resolved. You're instructed to check specific items on your preflight inspection for a reason, and if you can't do the check, then you shouldn't fly the airplane.

Case 7: Personal Experience—Loose Fuel Cap

I was flying a Beech Baron for a highway construction company. My mission that day was to fly to Nashville, Tennessee, pick up two passengers, and fly them to Johnson City, Tennessee, about an hour away.

It was early on a Saturday morning, and it was cold. I wanted to get fuel at Nashville, but they couldn't get the fuel truck started. Inside the fixed base operator (FBO), I quickly calculated that I had plenty of fuel for the Johnson City flight. My passengers arrived; we hustled out to the airplane, across the cold ramp. Then we strapped in, and I ran through the Before Start and Starting checklists.

I picked up my clearance and took off. Only after we were airborne did a passenger lean up from the Baron's rear seats to ask, "Did you know there's fuel coming out of the wing?" I glanced to the left side and immediately saw that the Baron's fuel cap was not tightened on, but instead it was sitting at a slight angle in the filler port. A thin stream of avgas was being sucked out by the low pressure created on top of the flying wing.

With no idea how much fuel had vented overboard and knowing that the float-type fuel quantity indicators in most light airplanes will indicate abnormally high levels with an unported fuel tank, I had no choice but to request an immediate landing to resecure the cap. We landed. I shut down and closed the filler port, then let the wing-mounted fuel quantity indicator settle down long enough for me to get a reading I was comfortable with. I still had plenty of fuel for the trip to Johnson City, so I got a new clearance and made the rest of the flight uneventfully.

Preparation

Obviously, the "line boy" at Nashville had loosened the fuel cap before going after the fuel truck. When the fuel truck wouldn't start he did not go back to the Baron, and so he did not resecure the cap. I was in such a hurry because of the cold weather that I did not do my own standard operating procedure of a quick walk-around, even if I'd just landed moments before, so I didn't catch the loose fuel cap either.

In-flight

I was busy climbing out from an international airport in instrument meteorological conditions. Still, I should have noticed the loose fuel cap and streaming fuel before my passengers. Once my passenger spoke up and I detected the problem, I got the airplane back on the ground, confirmed the remaining fuel level and determined it was still acceptable for the flight, and made the rest of the trip with no further incident.

Post-flight analysis

This minor incident, which may had led to a situation similar to the training twin in Case 5 if it had gone uncorrected, reinforced the need to follow standard operating procedures even when rushed by passengers or the extreme cold. It also prompted me to include a quick glance to the wings before starting up, and again as soon as possible during climbout, to recheck that the fuel caps are secure.

Lessons learned

Stated simply, flying is the culmination of a lot of human activity. There's opportunity for oversight and error at every step along the way. Standard operating procedures and checklists are our best defense against the natural results of the interaction of airplane and human.

Avoiding Fuel-Related Accidents

Printed checklists and standard operating procedures do wonders to compensate for the human factors of stress and distraction. They remind us to complete flight-critical items that most times are not forgotten—but may without warning go undone if the pilot is fatigued, neglectful, or complacent. Perhaps nothing is more blatantly "wrong" than a pilot running out of fuel. SOPs and checklist use may help prevent fuel starvation and fuel exhaustion.

2

Improper Takeoff Configuration

The most common cause of takeoff accidents in air carrier airplanes is the failure of the crew to correctly configure the airplane for takeoff. Not getting the airplane fully ready for flight is almost always the result of poor checklist use or distraction that causes an aircrew to omit critical flight items.

Although most general aviation airplanes are more forgiving than airliners when not precisely set up for takeoff, in demanding or unusual cases (such as a short field or a high-density altitude takeoff), it may still be necessary to fly precisely in accordance with pilot's operating handbook–recommended techniques in order to glean maximum performance from the airplane. And as we'll see in some of the Aviation Safety Reporting System (ASRS) accounts in this chapter, sometimes the preflight omission and the hazard to safe flight has nothing to do with the make and model of airplane involved.

Sometimes the checklist step missed, and the corresponding takeoff misconfiguration, is not itself crucial to the immediate safety of flight. Such omissions, however, are a warning of the *potential* for more safety-critical results from poor checklist discipline.

Case 1: Fasten Seat Belts (ASRS Accession Number 40770)

The crew of a Fokker 100 twin-engine jet was accomplishing the Before Starting Engines checklist when a gate agent brought a "jump seat" passenger into the cockpit. The ASRS reporting pilot, the captain of the Fokker, "allowed the checklist procedure to be interrupted" by the distraction, and the crew missed a required Before Takeoff checklist step—turning on the cabin Fasten Seat Belts signs. During the taxi-out the captain happened to notice that the seat belt sign switch was in the "off" position; he "immediately turned on the switch," and the flight continued normally.

This was a minor transgression, one that likely would not ever had led to an injury had the captain not noticed his oversight. But as the reporting pilot notes to the ASRS, "I now realize the importance of always maintaining normal procedures at all times. If something as seemingly obvious as turning on the 'fasten seat belts' signs before takeoff could be so easily overlooked, then what other, more critical items might be missed?"

Preparation

As we saw several times with fuel-related incidents, we again have a cockpit distraction during the processing of the Before Start checklist. Pilots should remember that if a checklist is interrupted for any reason, they should *positively* mark their place on the checklist before dealing with the distraction, and then return to the checklist at that point when they are able. If a checklist is interrupted and the pilot cannot determine where he or she was at the time of the distraction, the pilot should start the checklist over from the very beginning to catch any critical flight items.

In-flight

There was no in-flight phase in this occurrence.

Post-flight analysis

As the reporting pilot critiques, "I now realize the importance of always maintaining normal procedures at all times." This time it was a simple item, the Fasten Seat Belts sign—an important safety precaution to be sure, but not one that would impact the handling of the airplane. It could, however, just as easily have been the flaps, or the trim, or some truly crucial safety-of-flight item. From this ASRS narrative the crew and we have all learned a valuable lesson with little actual risk.

Lessons learned

It's easy to forget things when you're distracted. If you're alone in crewing the airplane, you have to be your own quality control manager. If something interrupts your checklist, start it over from the top. You don't need to redo tasks you *know* you've already done (like cycling fuel selectors or checking the magnetos), but you'll be certain to catch anything you missed as a result of the distraction.

Case 2: Takeoff without Clearance (ASRS Accession Number 409427)

An MD-80 crew was lined up for departure from Dallas–Fort Worth and received tower takeoff clearance. Just as the twin-engine jetliner lifted off, the first officer (FO) advised the captain that he had forgotten to request and obtain air traffic control (ATC) clearance for instrument flight to the destination. The crew maintained visual flight and quickly got clearance in the air. The crew also discovered that it took off with the transponders turned off.

Preparation

Picking up an instrument clearance and turning on the transponders before takeoff are both standard operating procedures (SOPs) followed by pilots, air carrier or otherwise. Good checklist discipline and clear divisions of responsibilities with cross-checks in the case of a multi-pilot crew should help pilots avoid this type of blunder.

Were the weather conditions worse at the time of takeoff, the pilots would likely have lost their certificates and their jobs and may have subjected their passengers and the occupants of other airplanes to collision hazards. This breakdown of procedure and checklist use might have had very serious ramifications if conditions had been just slightly different.

In-flight

Once the crew turned on the transponders and obtained instrument flight rules (IFR) clearance, what do you suppose was the first thing it did? What would you do? Most likely, the captain would have made sure the MD-80 was stabilized in a departure climb and then *pulled out the Before Start and any other necessary checklists* from the very beginning and ran through them with the first officer, step-by-step, to make sure nothing else had been forgotten. It's what any sane person likely would have done. The crew could have avoided its transgression by running the checklists with the same diligence *at the appropriate times*, instead of after-the-fact as the jet winged toward destination. As someone once said,

"There's always enough time to do things right the *second* time."

Post-flight analysis

The MD-80 captain, who authored the ASRS report, states that he "should have checked (for the clearance) twice" using the airline's standard operating procedures and checklists, and failed to do so. What the reporting pilot did not say is that often the familiarity of flight, especially if the trip is not the first flight of the day, may lull pilots into believing they've performed tasks they have not. The trim check, the fuel load, or even the instrument clearance you remember getting may not have been from that leg but from the one before. Checklists and standard operating procedures are designed *exactly* to combat complacency that familiarity and comfort with the airplane can bring.

Lessons learned

Again, there's no one but you if you're alone in the cockpit. It should be easy to remember to obtain your clearance. However, if you're flying multiple legs, if you're tired, or if you're flying a very familiar route, be especially careful to check *everything* before you go.

Case 3: Another Clearance Missed (ASRS Accession Number 411830)

Airliner crews often receive their clearance through a printer in the cockpit called the ACARS. A Boeing 727 crew, preparing to depart Miami, Florida, found the ACARS printer to be inoperative, a "deferred maintenance" item. Now faced with a deviation from standard operating procedure, having to get ATC clearance over the radio instead of through "normal" means, the crew was also distracted by ground crews, caterers, and the other myriad details of launching a jet full of passengers. The crew forgot to request clearance.

On takeoff, Miami Tower assigned the flight a transponder code, and only after checking in with departure control did the aircrew find it did not have a clearance. Departure control radioed clearance, and the flight continued uneventfully to destination.

Preparation

Often there are clues to upcoming transgressions. In this case, the first clue was the inoperative ACARS. That alone should have alerted the crew to be especially watchful for problems associated with ATC clearance. The second clue, actually a series of clues, was the numerous distractions that seemed to be sufficiently greater than normal that they warranted mention in the ASRS report. Third, Miami Tower's assignment of a discrete transponder code should have "rung a bell" with the captain and crew. Normally, transponder codes are received as part of an instrument clearance, so receiving a code from Tower by definition means that the crew had not received or was not following an assigned clearance.

In-flight

In flight, the crew (which at that point did not have a defined route or altitude to fly to destination) received clearance, and everything else was uneventful. Miami Tower, which saw a jetliner preparing to take off that did not have a clearance, and which assigned a transponder code to the flight without querying the crew, should share some responsibility for the transgression. In the air, though, the system worked because Departure obtained and radioed the clearance, and the flight went on as planned.

Post-flight analysis

The reporting pilot in this case did not offer any critique or evaluation, nor did he make any recommendations about changes in habit or procedure to avoid a similar occurrence in the future. The pilot did, however, mention that the distractions caused the crew to miss items on the Before Start checklist, which led to taking off without a clearance. All pilots flying as part of an airline crew or as a single pilot in a general aviation airplane should be alert to distractions and be especially careful if changes in standard operating procedures (like the method of picking up a clearance in this incident) become necessary.

Lessons learned

Beware of abnormalities—they're warning signs. In this case, the crew received a transponder code from the tower, just before takeoff. That's not normal when leaving a controlled airport; normally you'll get your transponder code from clearance delivery or ground control before you ever leave the ramp. An unusual delivery of information is a clue that something isn't going as planned.

Watch also for any time that technology or other factors make you change your standard operating procedure. In the case cited above, the crew had to get instrument clearance through what for it was alternative means. If you find you need to do something differently from what is your norm, remember that "different" means "more likely to be done incompletely or incorrectly." Take your time and use your checklist as a backup.

Case 4: Failure to Configure for Takeoff—Lights (ASRS Accession Number 386580)

The pilot of a Cessna 172, who had owned and flown the airplane for 18 years, taxied for a night departure. He diligently completed the Before Takeoff checklist at the end of the runway—or so he thought. Climbing out onto a downwind leg, he "found that I had taxied and departed with a landing light only, with no navigational lights or [rotating] beacon." He continues: "I always use the beacon, day or night, so I have no idea how I missed 'Lights as Required' on the checklist." The pilot turned on the appropriate lights and continued without incident.

Preparation

For most pilots, night flying is an unusual occurrence. Anything that's out of the ordinary should cue a pilot to be especially careful not to miss checklist items. Fatigue is also often a factor in night-flying incidents, so pilots need to remember that fatigue may cloud judgment and affect actions; strict checklist discipline helps greatly to compensate for the effects of pilot fatigue.

In-flight

The pilot in this case noticed his omission while on the downwind leg and simply turned on the appropriate lights.

Post-flight analysis

The reporting pilot offers this critique: "The [Air Traffic Control] Tower never said a word [about the lack of position lights], and fortunately was not working any other aircraft" at the airline-served airport. "In the future, in addition to completing the checklist and rechecking on climbout, I'll give the panel a 'once over' as I take the runway."

Lessons learned

If you're doing something unusual for you (flying at night, flying an unfamiliar airplane or route, and so on), remember that "unusual" means "unpracticed," and more than ever you need to follow good procedure and use checklists to avoid any oversights.

Case 5: Failure to Configure for Takeoff—Flaps (ASRS Accession Number 402337)

A Boeing 757 lined up for takeoff and the "pilot flying" advanced the power levers. As the engines spooled up, a "Configuration Warning" sounded in the cockpit, and the crew reduced power, taxiing clear of the runway. The flaps and slots were still in the "up" position; if the crew had attempted takeoff without the aerodynamic aid of these lift-enhancing devices, the aircraft might well have not been able to clear obstacles on takeoff, with disastrous consequences.

As the reporting pilot puts it, "We were distracted during those portions of the taxi [from the gate to the runway] when the Takeoff checklist is normally performed, and once outside that window we failed to notice our omission." The captain further counts a radio discussion about weight and balance for the flight, an unusual taxi route, and a significant delay before takeoff as factors leading to the oversight.

Preparation

Again, we have "unusual" events cited as contributors to a checklist procedure oversight and a deviation from standard procedures (a different taxi route, the need to perform the Takeoff checklist at an "other-than-normal" point during taxi) leading to what might have been tragic. Only enhanced cockpit technology, in the guise of the Configuration Warning alarm, prevented what could have been a horrific accident and tremendous potential loss of life. Earlier models of airplane didn't have this kind of warning; it exists because there *have* been cases where distraction led to attempted takeoff in an incorrect airplane configuration.

In-flight

There was no "in-flight" portion of this trip on which to comment, thanks to the engineering advancement of the B757 over earlier models of airliners.

Post-flight analysis

This is simply another case where unusual procedures and distractions led to missed checklist steps, and but for advanced configuration monitoring and warning systems, those missed steps quite likely would have led to disaster. All pilots need to be aware of the effects of distractions and deviations from the norm; having to deal with interruptions or unusual procedures in themselves should be warning signs for pilots to check and recheck critical flight items and rerun checklists, looking for omissions.

Lessons learned

Here's another case of running checklists while taxiing, which may be acceptable with more than one crew member but which is an invitation to disaster for a pilot flying alone. Note how even a professional, two-pilot crew missed a critical flight item by performing checklists while taxiing. Your airplane quite likely is not as sensitive to flap position on takeoff as is a modern jet-liner, but there are items (for instance, trim setting) that can bring a small airplane down if set incorrectly for takeoff. Unlike the 757, you'll usually not have a warning indicator if you try to take off in a misconfigured airplane. Take enough time to be sure everything's set correctly before beginning your takeoff roll.

Case 6: Cabin Pressure Misconfiguration (ASRS Accession Number 403139)

Thunderstorms ringed the area around Denver as the 737 took off on a night flight to San Francisco. The captain was the "pilot flying" while the first officer handled communications and miscellaneous chores on this leg. The FO received and answered several radio calls on the company channel as the jet climbed out of Denver.

Passing through about 20,000 feet, the crew's Cabin Pressurization alarm sounded. Silencing the alarm, the FO discovered that the cabin altitude was reading 10,000 feet and climbing—the pressurization system was not holding cabin altitude below 10,000 feet as required by Federal Air Regulations Part 121.

The abnormal procedure called for the crew to don oxygen masks before troubleshooting, which it did. On oxygen, the crew focused attention on its tasks—the captain was still primarily flying the airplane and requested a lower altitude from the Air Route Traffic Control Center, while the first officer tried to fix the pressurization problem.

Descending through 14,000 feet, the panel alerted the crew that oxygen masks had automatically deployed in the passenger cabin. The cabin pressure had topped

14,000 feet; the 737 was flying without cabin pressurization. The captain received clearance back to Denver. During the turn back toward Denver, the captain discovered that the FO had not switched the pressurization controls on after flap retraction, which is part of the standard Climb checklist procedure. Since the passenger masks had dropped, however, the crew was required to abort the trip and return to the point of origin. The 737 was grounded until the passenger emergency oxygen masks were reinstalled, and the crew completed the flight much later on a replacement airplane.

Preparation

The crew appeared to be following current-standard airline procedures for division of attention, one pilot focusing on the direct flying chores and the other handling miscellaneous items as they came up. The thunderstorm activity in the area required the first officer to be actively working with the radar, while the company-channel radio calls interrupted his flow as well during the climbout. The FO missed a single, critical checklist step, and because of the "pilot flying/pilot not flying" procedure, the captain did not cross-check and discover the omission before it led to the extreme cost and public relations nightmare of an oxygen-mask-required return to the airport. Better division of attention, but with cross-checks of each other's work and better adherence to checklists, would have prevented this costly and embarrassing mistake.

In-flight

Dealing with the indications en route, the crew still failed to use checklists to determine the true nature of the problem until it was too late and the trip was required to be aborted. "We focused all our attention on the pressurization panel," reports the first officer, who filed the ASRS report. The crew was treating the symptom, the loss of cabin pressure, and not the cause, the position of the engine bleed air system controls that provide cabin pressurization in the first place.

Once faced with the mandated return to Denver, the crew admirably chose to return immediately, and got its passengers back on the ground without further incident.

Post-flight analysis

There were a lot of contributing factors in this case. First, fatigue: This was a night flight, the last leg of the first day of the crew's multiday schedule. The first officer, who lives in Michigan and commutes to his Denver base from Detroit, has "previously shown poor performance near the end of the [first duty] day." In his self-critique the FO recommends he should commute to Denver the day before his flight schedule begins instead of on that morning. Further, the crew had a three-and-a-half-hour layover at Denver between its last trip and this flight. Multiple distractions during the initial stages of the flight compounded the scenario. "Because I allowed myself to get caught up in [working] the radios and possibly [because of] fatigue," reports the first officer, "I forgot to finish the checklist. In other words, poor radio and checklist management."

The FO also cites the physical factors of dark night flight conditions, a "dimly lit" overhead control panel, thunderstorm avoidance, and "occupied time" as contributors.

The flight's captain, too, filed his own ASRS paperwork. His description of the events closely matches that of the first officer and he has these criticisms to add:

"Company policy states that all takeoffs out of Denver are done with the engine bleed [controls] off. At 10,000 feet AGL [I] called for the climb checklist," just when "we were told by the tower to switch over to departure control. This caused [us] to stop what we were doing in order to respond to the tower controller and [to] call departure control. By the time [we] were done talking to departure control we were at 10,000 feet, the altitude when [we] normally call the company" to report departure times. On the way back to Denver the captain "realized [we] did not finish the Climb checklist because resetting the engine bleeds to the 'ON' position is an item on the Climb checklist."

Lessons learned

Checklists are designed specifically to help you overcome the possibility of forgetting something, a possibility that is heightened in a busy environment or when you are tired. But checklists can only help if they're used. Especially if you're tired or if you're faced with multiple distractions, use your checklists. If you're faced with an abnormal indication, like the 737's pressurization, don't depend on your memory alone to deal with the problem. You likely have printed checklists for most unusual situations; fly the airplane, get the indications under control, then reference the checklist to figure out what else you can do to remedy the situation.

Case 7: Canopy Open on Takeoff (ASRS Accession Number 389421)

The sleek Extra 300 aerobatic airplane was departing from Santa Barbara, California. The pilot obtained his clearance out of the Class C airspace, writing his departure clearance on a piece of paper which "covered [his] checklist." With his checklist reminder obscured, "after run-up and taxi I neglected to check that the canopy was down and locked.

"About two seconds after takeoff the canopy opened and broke against the wing." First, the pilot attempted a landing on the remaining runway, but soon it was apparent there was not enough runway ahead of the Extra for the pilot to make a safe landing, and he applied power for a go-around instead. Crosswind and downwind legs were without radio communications with the tower "because of the way my headset was blown off by the airspeed and prop wash." The pilot made a successful landing without radio communication with the tower.

Preparation

As is often the case with "bubble canopy" airplane designs, it was hot in the cockpit. Accepted procedure is to taxi such airplanes with the canopy open, or at least not completely secure, so air can flow into the cockpit. The pilot must remember to latch the canopy completely secure before taking off. The Before Takeoff checklist helps the pilot remember. But by his admission in the ASRS report, the pilot in this case covered up his Before Takeoff checklist with a piece of paper on which he wrote his departure clearance. Better cockpit organization, including a place for his written clearances that does not cover up critical checklist items, might prevent similar occurrences in the future.

In-flight

Once aloft and faced with the enormous distraction and noise of the slipstream, the pilot seems to have performed well. He maintained control of the airplane and at first tried a straight-ahead landing. When it became obvious that wouldn't work, he went round, only to find that he had no way to communicate with the tower or to obtain a landing clearance by radio. Regardless, he maintained positive control of the airplane and of the flight, remained in the traffic pattern, and made a successful landing.

Post-flight analysis

The reporting pilot attributes his experience to a "chain of events: canopy unsecured due to inadequate [Before Takeoff checks] during run-up. Contributing factors include heat in the cockpit and the checklist being covered by a paper on which the clearance was written."

Rightfully, the pilot notes that "other than the initial incomplete/inadequate [Before Takeoff] checklist, pilot response seems adequate." He does observe, however, that "perhaps some indicator other than [canopy] handle position could be installed to verify the canopy being locked," since there are "no observable elements [to] indicate canopy security." The reporting pilot calls his checklist deviation "a real learning experience."

Lessons learned

Standard operating procedures, good cockpit organization, and perhaps a just-before-takeoff "final items" checklist can help you be certain the airplane is truly ready to fly.

Case 8: Propeller Strike on Takeoff (ASRS Accession Number 389796)

Lest you think a pilot's attention should be focused *completely* on a checklist before takeoff, read the experience of this commuter airline crew:

The crew of a Jetstream 31 twin-engine turboprop was taxiing out for a night departure at Burbank, California. The captain was maneuvering the airplane on the ground while the first officer was helping with the Before Takeoff checklist. "Slightly distracted by correcting the First Officer's use of nonstandard calls during the checklist," reports the captain, "I inadvertently lined up on the left edge line instead of the centerline" of the departure runway.

"During the initial takeoff roll I felt and heard a small bump, which I thought was the aircraft's left main wheel passing over an embedded surface light. At that time I realized we were on the left edge of the runway, and [I] corrected toward the runway centerline. Because I thought we had only run over an embedded runway light, dividing the taxiway from the runway, I continued the flight to the scheduled destination. After an uneventful flight and landing I discovered damage to one of the left propeller blades. Apparently the propeller had struck an object on the ground, possibly a runway light."

Preparation

This incident might have been avoided if the crew had better reviewed the runway lighting information for the airport, which showed that the runway in use did not have centerline lighting. A more disciplined cockpit procedure during taxi-out might have provided better opportunity for both crew members to scan outside while running the checklist.

In-flight

Once the Jetstream was airborne, abnormal indications ceased and the crew continued with no knowledge of the damage. The crew did not have to change its operation of the twin turboprop because of the incident.

Post-flight analysis

The crew failed to monitor outside of the airplane while setting up for takeoff. The captain, who filed the ASRS report, cites "the lack of the first officer's attention outside the aircraft while completing the Before Takeoff checklist." The captain also wishes the first officer had been more vocal and observant: "He [the FO] was silent throughout the takeoff and did not notice our close proximity to the left edge of the runway until I told him I was correcting back to runway center."

Lessons learned

Running checklists while taxiing greatly diminishes a pilot's ability to see and avoid obstacles. It encourages rushed action, can cause the pilot to miss critical flight items, and may well be a factor in runway incursions and other mishaps. If a pilot is in so much of a hurry that he or she cannot stop the airplane long enough to complete the Before Takeoff and other checklists, that pilot should take the time-stress itself as a sign that he or she is more likely to forget something. Slow down and get it right.

Navigating on a large airport with complex systems of taxiways and runways takes concentration, especially at night, in low visibility, or if the pilot is not familiar with the field. Take a moment to study airport diagrams before starting the airplane, and brief yourself on the type and location of lighting to expect on the departure runway. This might help you avoid experiences like that of the Jetstream crew, as well as more serious runway incursions or even attempted takeoff from the wrong runway.

Case 9: Two Engines—Why Not Use *Both*? (ASRS Accession Number 445237)

A newly hired first officer was receiving his Initial Operational Evaluation, or IOE, in a Fokker 100 twin-jet airliner. On the sixth flight of the day, as the check pilot and author of the ASRS account reports, "We were distracted during taxi discussing the special [arrival] procedures for our destination. We had discussed these at the gate and the discussion continued during taxi. Due to this [distraction], we failed to perform the required checklists after start and [during] taxi.

"When we arrived at the departure end of Runway 36 Center, we were cleared into position and then cleared for takeoff. As we made the turn to align with the runway, I noticed we had not completed the checklists and the Number 1 engine was not started." The crew requested and was granted taxi clearance down the runway to the first turnoff, exited the runway, and completed the checklists before proceeding back to the runway and completing an uneventful flight.

Preparation

Obviously, this crew needed to better coordinate its actions and cross-check each other. Perhaps the check captain wanted to see how far the new first officer would go without running checklists. A more likely scenario, however, was simply that the check captain diverted his crew's attention with an evaluation of the special procedures at the destination.

In-flight

There was no in-flight phase in this incident.

Post-flight analysis

Writes the reporting check captain, "We were tired and hot from a long day, and failed to give due diligence to checklist completion, including starting the [left] engine."

Lessons learned

As the reporting check captain succinctly observed, "When tired, be careful."

Case 10: Improper Trim Setting (NTSB Report Number DEN00LA043)

A high-performance airplane collided with terrain during its initial climb from an airport in Utah. Luckily, the private pilot and his three passengers were not hurt. The airplane, however, suffered substantial damage.

"According to the pilot," states the National Transportation Safety Board (NTSB) preliminary report, "'during the takeoff roll the aircraft was trimmed incorrectly and [it] left the runway prematurely.' The airplane lifted off the runway, settled back down, and immediately lifted off the runway again. The aircraft has insufficient speed and the stall warning horn sounded. [The pilot] retarded the throttle and the airplane departed the runway to the left into the grass. The propeller, cowling, and landing gear were bent, and the fuselage was wrinkled."

Preparation

Some types of airplanes are very sensitive to pitch trim setting. The heavier the load, typically, the farther aft the airplane's loaded center of gravity (c.g.), and (again typically) the farther rearward the c.g., the less stable the airplane. Furthermore, some airplanes' landing pitch trim position is significantly different from where it needs to be for a safe takeoff; taking off with a mis-set pitch trim may introduce aerodynamic forces that are difficult if not impossible for a pilot to overcome in an already unstable airplane.

Pilots are not often educated on the significance of weight and balance when moving up to a high-performance airplane, and they rarely have the chance to practice, under controlled conditions and with a seasoned instructor pilot, controlling the airplane in radically out-of-trim configurations. Many pilots also begin flying high-performance airplanes without a solid foundation in the need for good checklist use, which likely would have allowed this pilot to avoid a dangerous, uncontrolled flight.

In-flight

Once airborne, this pilot and his passengers were along for the ride, unless he had been able to firmly overpower the out-of-trim controls while aggressively dialing out the excessive trim deflection.

Post-flight analysis

To quote the NTSB preliminary report, "When asked what recommendation the pilot could make as to how the accident could have been prevented, [the pilot] stated 'use of a checklist.'"

Lessons learned

One little omission can be enough to bring down an airplane and its occupants. Before takeoff, pull out the checklist (or at least work through a mnemonic procedure) to make sure you haven't missed something crucial.

Avoiding Improper Takeoff Configuration Accidents

Takeoff is historically one of the riskiest phases of flight. Airspeed and altitude are both low, and there's little time to recover if something was not properly configured for takeoff. If you're distracted, or forced by some outside reason to change your operating procedure for a flight, use that knowledge as a reminder to concentrate even more on getting everything done right. Checklists and proper use of standard operating procedures will help you accomplish the critical flight items before takeoff.

3

Improper Landing Configuration

Like improper takeoffs, landings, too, can be marred by poor checklist use or failure to follow standard operating procedures (SOPs).

Case 1: Hard Landing (ASRS Accession Number 386703)

A cargo DC-9 was descending into Wilmington, Ohio. The first officer was flying the airplane; the captain, acting as the "pilot not flying," was responsible for completing the Before Landing checks. "During the Before Landing checklist," reports the captain to NASA's Aviation Safety Reporting System (ASRS), "I failed to arm the ground spoilers." Spoilers are part of the wing drag system that allows a jetliner to stop in a relatively short distance on landing.

"On short final," continues the captain, "I noticed that the ground spoilers had not been armed. As the first officer was reducing power to idle (for the flare), I reached down and instead of arming the spoilers, I deployed them. This resulted in a hard landing with damage to the tail section of the plane." The captain reported the hard landing to company maintenance technicians, who on inspection found "extensive damage" to the DC-9.

Preparation
Better procedural and checklist discipline would likely have prevented this mishap, which could have been far worse.

In-flight
The captain in this instance exhibited the personality trait of impulsiveness—he saw a discrepancy and immediately moved to fix it, but without taking the time necessary to ensure that his action was indeed correct for the situation. It likely would have not mattered at all to delay arming the spoilers for the few extra seconds it took to properly identify the control and verify it was moved correctly.

Post-flight analysis
The reporting pilot cites distractions as contributing factors in his omission. "The time leading up to this incident [was] routine with the exception of a runway change 15 miles from the airport," he writes. "Inside the outer marker it was necessary for me to call Dayton Approach to ask for a change to the tower frequency. This Dayton Approach Control contact interrupted my Before Landing checklist. It was during this time, in my opinion, that I failed to arm the spoilers."

Lessons learned
Beware of distractions that may cause you to forget checklist items. Be even more wary if you find yourself making impulsive actions, performing tasks before com-

pletely thinking them through or determining that your action is correct for the conditions. One function of checklists is to help properly pace and sequence actions.

A "go-around" to set things right might have seemed costly or excessive before the tail hit the ground, but it now seems economical and correct given the outcome. If something's out of place on short final, don't try to fix it in flight; apply power, smoothly go around, and then take the time to correct the problem and follow up by referencing the checklist.

Case 2: Landing without a Clearance (ASRS Accession Number 393080)

A McDonnell-Douglas Super 80 was on approach to Houston, Texas. The approach controllers were very busy; the airliner was not given a descent from 10,000 feet until it was within a few miles of the airport. Finally, Approach cleared the flight to 4000 feet and asked the crew to report when it had the airport in sight.

The crew picked up the runway visually and called Approach; it was cleared for a visual approach, but was "preoccupied with getting the airplane 'in the slot' for a successful landing." The reporting pilot later writes in the ASRS that "we were late getting the checklist done." Further, the crew was using a checklist newly adopted by its airline, a checklist that omitted steps verifying the crew had been "cleared for the approach" and "cleared to land." Being very busy and not prompted by an otherwise properly run Before Landing checklist, the crew landed the airplane at Houston without being cleared to land by the air traffic control tower.

After landing, ground control asked the crew if it had in fact been cleared to land. The reporting pilot replied affirmatively but was uncertain enough that, once parked at the gate, he called controllers on the phone and was told the tower had in fact flashed the flight a "green light," clearing the crew to land. "There was no conflict," according to the report.

Preparation

This is a case where a company-mandated change in checklist format and standard operating procedure was adopted, but where the crew was not properly trained to the point where the new procedure would be automatic. Further, for some reason, the company felt it did not need to include confirmation of what are, of course, two critical clearances (to approach and to land) on the new checklists. Crews had depended on the checklist to remind them in the past and now no longer had that reminder. More practice in the new procedure and some replacement means of verifying clearance to land might have prevented what could have been a dangerous traffic conflict.

In-flight

Every indication was that the crew was properly using the checklist made available to it. It was the failure of the checklist authors to include items the crew deemed necessary that led to this incident.

Post-flight analysis

The reporting pilot calls this instance "the same old story of getting distracted [by the close-in, rapid descent] in the cockpit. I also feel," continues the pilot, "that if we had the final check 'cleared to land' as we did on the old checklist, this omission would not have occurred. Right now I'm searching for something else I can use in the cockpit to replace those missing items on the [new] checklist."

Lessons learned

Changes in procedure require practice to become "natural." If you're changing the type of airplane you fly or the manner in which you fly it, be especially careful to review checklists and procedures ahead of time to make sure there's nothing you're likely to miss. Try to incorporate changes gradually so you're not completely overwhelmed with "new" ways of doing things to the point you forget a critical flight item.

Of course, if you're a Part 91 operator, you shouldn't have to accept checklist or procedural changes unless you want to. Remember, though, that any change in standard ways of doing things will be inherently riskier until the new way becomes second nature.

Case 3: No Steering on Landing (ASRS Accession Number 384765)

A Boeing 757 crew lined up for landing and extended the landing gear. Continuing inbound, the crew checked the gear indications and discovered that the nose landing gear indicated "unsafe." After completing the Alternate Gear Extension checklist with no change in indications, the pilots declared an emergency and made a normal landing, "rolling straight ahead."

The gear was "fully down and no damage occurred." But there was no nosewheel steering available, and the 757 had to be towed from the runway. On reflection and investigation, the reporting crew member realized "when completing the Alternate Gear Extension checklist, we left out the step of putting the (landing) gear handle to 'down'—it was left in the 'off' position. This resulted in no nosewheel steering on landing. Once [the] gear lever was placed 'down,' all systems functioned okay—except [that the] nose gear light was still out due to a bad [light] socket."

Preparation
The flight crew in this case quickly identified the problem but was hurried to find a solution while continuing on the approach and landing. As it was, the crew's impatience did not seriously endanger the passengers or itself, but in a strong crosswind, or any number of other possible scenarios, the lack of steering capability could have proven disastrous. The crew might have realized the possible danger of trying to fix a gear problem while on final approach, climbed to a safe altitude and on a vector during which it could run the checklist under less stress, and then returned for a normal landing.

In-flight
Of course, airline crews are under a great deal of pressure to land as soon as possible. It's extremely costly, and seen as bad customer service, to miss an approach and delay completing a flight. The crew's stress, however, caused it to land an airplane with seriously degraded control authority, endangering passengers and itself. Obviously, the crew thought it was hazardous enough (to passengers and/or its livelihood) that it filed Aviation Safety Reporting System paperwork to report the problem and to gain regulatory immunity.

Post-flight analysis
"I feel my crew resource management was poor and I should not have missed any checklist items," writes the reporting pilot in his self-critique. "I allowed myself to rush."

Lessons learned

Don't allow yourself to be rushed through a procedure. Especially avoid trying to run an emergency checklist while at the same time flying an instrument or visual approach. In almost all cases, you'll be far, far ahead to break off your approach, climb to a safe altitude to troubleshoot and resolve the crisis, and only then return for another landing attempt. With just a few more seconds to work through an abnormal procedure, you'll be far less likely to miss something critical. And with time to read through the checklist, you'll catch anything you might miss under the demands of an abnormal or emergency procedure.

Case 4: Should've Kept the Wheels Up (NTSB Report Number SEA00WA064)

A charter pilot, a cabin attendant, and seven passengers escaped injury in the crash of a Cessna 208 Caravan amphibian while landing at the Filitheyo Seaplane Base, Dhiguyvaru Falhu, in the Maldives.

According to the NTSB, the pilot reported he was "distracted by a radio conversation with Air Traffic Control" while departing Male International Airport for the 30-minute flight to Filitheyo. Because of the distraction, he forgot to raise the amphibian's wheeled landing gear. En route, he "did not notice any vibration or turbulence" related to the extended wheels, "and [he] was unaware of any reduction in normal cruise speed. As he approached the Filitheyo Water Aerodrome, the pilot realized that the coordinates that [he had] set in the GPS for the landing area were actually the coordinates for the resort island of Filitheyo, about 2.5 miles north of the landing lagoon. The pilot therefore became busy resetting the correct coordinates into the GPS and did not notice that the [wheeled landing] gear was still extended.

"After touchdown, the extended landing gear dug into the water, and the aircraft flipped over on its back. The pilot and cabin crewperson helped the passengers exit the aircraft, and all occupants were picked up by a transfer boat."

Preparation

Better familiarity with amphibious Caravan procedures, preflight planning that included determining the correct Global Positioning System (GPS) coordinates for his landing spot, and better prioritization and cross-checking in flight, including the use of Climb, Cruise, and Before Water Landing checklists, would likely have prevented this pilot from risking the lives of his passengers and seriously damaging a million-dollar airplane.

In-flight

The pilot reports he was "unaware of any reduction in normal cruise speed" created by flying with the landing gear down. However, careful calculation of expected cruise performance for a given power setting and set of environmental conditions and close observation of what was actually obtained in cruise likely would have revealed at least some noticeable difference that might have clued the pilot in to awareness that the gear was down. And again, referencing any number of in-flight checklists while en route should have prompted the pilot to raise the amphibious gear before attempting a water landing.

Post-flight analysis

Checklist use and a standard operating procedure that includes predicting airplane performance and comparing that prediction to actual figures would likely have prevented this mishap.

Lessons learned

This is yet more proof that using checklists as reminders can help you catch those "obvious" items that sometimes get forgotten in the workload and stress of flight.

Case 5: Setup for Disaster (NTSB Report Number FTW01FA015)

A student pilot and his flight instructor died in the aborted landing of a Beech 76 Duchess. The twin-engine training airplane was destroyed in the crash.

The instructor pilot radioed Tulsa, Oklahoma, Approach Control from a position about four miles from the Okmulgee Airport, stating the engine was on fire. The pilot requested a straight-in landing on Runway 1 Left at Riverside Airport in Tulsa, the airplane's home base.

"The ATC personnel asked the pilot to describe the problem again," states the NTSB's preliminary report, "and the pilot reported an engine fire. The approach controller asked the pilot to indicate how many people were on board, how much fuel was on board, and which engine was on fire." The instructor answered, "Two people, two hours of fuel, and the right engine."

Riverside Tower controllers say they saw the airplane approach "fast" and saw "smoke coming from the right engine." One controller "reported seeing flames coming from the right engine.

"According to the [Federal Aviation Administration] inspector, the tower personnel reported that the [instructor] pilot stated he did not have a 'green light,'" apparently referring to the green light(s) which indicate that landing gear is down and locked. "The flight instructor then stated he was going around and was going to attempt to land on [Runway] 19 Right. The controllers stated that the airplane initiated a climb," then began to "roll over to the right and pitch nose down. The airplane impacted the ground and a fire erupted on impact. The tower controllers reported to the FAA inspector that the [Duchess'] landing gear appeared to be extended" prior to the crash.

"Numerous witnesses, located at various areas on the airport, reported that the airplane approached downwind to Runway 1 Left and was 'very fast on [the] approach.' The witnesses then heard an application of engine power and [saw] the airplane pitch nose up and [start] a climb. The witnesses stated that the airplane then slowed and rolled right [until it was] inverted, and pitched nose down until it impacted the ground. Some of the witnesses stated that the landing gear appeared to be in the extended position."

Preparation

Being inside an airplane burning in flight is one of the most deep-seated fears of most pilots. Certainly, coming in with a student at night and with an engine on fire, the instructor in command of this Duchess was under some of the greatest stress of his life.

Extreme stress tends to tap all our mental reserves, making us rely in large part on preprogrammed responses. It's unclear whether the pilots of this airplane feathered the right propeller or shut off fuel and electricity from the burning engine. The situation may have prevented electrical power from reaching the Duchess' landing gear lights, but for whatever reason, the instructor decided to go around. He lost control in the attempted single-engine go-around, however, and neither instructor nor student survived.

Better practice in engine-out, engine fire, and single-engine go-around drills, along with study of the landing gear extension and indication system might have prevented this tragedy.

In-flight
Faced with this in-flight emergency, the instructor had the opportunity to practice good cockpit management skills, which would have included using the student to help. We'll never know if the two worked together or if the instructor took command and the student was merely along for the ride. Cooperation between the two pilots aboard the Duchess might have resulted in better assessment of indications and airplane status and led to a much more desirable outcome.

Post-flight analysis
There's little hard data to go on after the fact in this case. For some reason, there was a discrepancy between what numerous observers on the ground saw (the landing gear extended) and what the instructor pilot recognized from the cockpit (lack of a "green light" gear-safe indication).

Lessons learned
An in-flight emergency of this severity can still be met by processing through a series of checklist procedures: Engine Fire in Flight, Engine Shutdown/Securing, Single-Engine Approach, Before Landing—Single Engine, or whatever the precise titles for the situation you may face. Pull out the checklists and run through scenarios such as the Duchess' final moments. Sit in the cockpit while you're doing so, and move your hands through the steps you'll complete for that scenario. Think not only about how to accomplish steps from a single checklist but also

envision the total scenario and think your way through everything you'd need to do to get the airplane on the ground. If you're faced with the very unlikely situation of flying an airplane under emergency conditions, wouldn't it be better to have thought the whole thing through ahead of time, before you were under the stress of reality?

Avoiding Improper Landing Configuration Accidents

Most aircraft accidents happen during the approach and landing phase. The airplane is converging with the surface at the same time the pilot is making rapid changes to the airplane's configuration, and the handling characteristics of the aircraft are constantly changing. There's little room for oversight or omission.

Practice your landing checklists and establish a personal standard operating procedure that will help you properly configure the airplane for landing and ensure you receive the landing clearance you need if necessary.

4

Miscellaneous Accidents and Incidents

Any number of other incidents can be attributed to improper checklist use, deviations from standard operating procedures (SOPs), or simply not having good habit patterns.

Case 1: Descent below Cleared Altitude (ASRS Accession Number 419420)

The crew of a commuter turboprop was in an "expedited descent" when the autopilot failed to capture the assigned altitude. The reporting pilot writes to the Aviation Safety Reporting System (ASRS) that "I noticed the descent [continuing below the] assigned altitude, but it took 300 feet [more of altitude loss] to stop the descent." The captain and first officer (FO) "were both distracted by finishing [the Before Landing] checklist and setting navaids. Someone always needs to be monitoring performance" even when flying on autopilot.

Preparation

A better-practiced division of "pilot flying" and "pilot not flying" duties might have prevented this altitude excursion. One pilot would have been setting and monitoring the autopilot and resulting airplane performance, while the other would have been processing the Before Landing checklist. A quick cross-check at some less critical time in the flight (*not* while leveling off from a descent) by the other pilot might have caught any discrepancies.

In-flight

The crew should have identified the time while it neared level-off from the descent as an instance where checklist action should be delayed. The crew could have reviewed previously accomplished checklist steps and completed the Before Landing checklist after establishing level flight on the assigned altitude.

Post-flight analysis

The reporting pilot puts it well when he writes, "Someone always needs to be monitoring performance." Autopilot use is not an excuse to stop "flying" the airplane.

Lessons learned

Don't depend on cockpit automation to fly the airplane without your help. It's your responsibility to properly engage autopilots and to keep a watchful eye to be sure the airplane is doing what you think it's supposed to be doing.

Remember, too, that checklists are designed to compensate for distractions, not to be distracting themselves. Whenever possible, delay checklist actions until the airplane is established in a stable flight condition (cruise, climb, or descent at a constant airspeed and trim setting); delay checklist steps if you're within, say, 1000 feet of your assigned or requested altitude so you won't miss leveling off. A little discipline here will go a long way toward keeping you safe and out of trouble.

Case 2: Continued Flight in an Unairworthy Airplane (ASRS Accession Number 403862)

The jet airliner began its takeoff roll on a trip from Newark, New Jersey, to Detroit, Michigan. The FO called rotation speed, and the captain, who was acting as the "pilot flying," rotated at a "normal" rate. The aircraft "continued to rotate," however, and the jet's tailskid struck the runway prior to liftoff.

Once established in climb, the FO got out the company's Non-Normal checklist and informed the captain of the established procedure: depressurize the cabin and return for landing. The captain, however, decided to press on to destination; the FO "made a very strong case for return but was overruled by the captain." The flight continued to Detroit.

After landing and deplaning the passengers, the captain and FO inspected the tailskid. They found damage; the captain "wanted to leave" without reporting it to maintenance or the company's flight department. The FO went over his captain's head and reported the unairworthy airplane.

Mechanics inspected the airplane and agreed that it was unsafe to fly. They advised the FO to "find the captain and force him" to report the incident. Unable to locate the captain, the FO was "forced to report the incident to the [airline's] Operations Director" personally to satisfy the company's procedure to get the airplane fixed.

Preparation

Airline crews spend a huge amount of time practicing and being evaluated on their ability to follow established standard operating procedures. With airplanes as complex as jet airliners, it's reasonable to expect pilots to need guidance in abnormal procedures, and airlines and aircraft manufacturers try to envision every possible scenario in order to provide aircrews with improved decision-making capability in the event something goes wrong. In effect, it allows airplane designers and airline managers to "participate" in decision making, allowing the crew the benefit of their insight and experience in a stressful time.

Although checklists and SOPs can't foresee every possible scenario, the crew did have a company procedure for dealing with a tail strike on takeoff, and it was simple—depressurize the cabin and return to the point of origin for landing and repair. The first officer, apparently well schooled in company checklists and procedure, referenced the checklist, advised the captain of its direction, and made a "strong case" for following the correct procedure. Instead, the captain (perhaps fearing discipline over the tail strike) overruled the FO and the airline's guidance and continued to destination. The decision was patently unsafe, put the passengers and crew at great risk, and eventually cost the captain his job.

Better evaluation of company stresses (for instance, the likely outcome of discipline applied for a tail-strike incident) and review of captains' decision-making ability under stress might have prevented this deviation from standard operating procedures.

In-flight

It must have been a tense situation in the cockpit, with the first officer trying unsuccessfully to convince his captain to return to Newark. The FO "did everything short of taking the controls" from the captain in an effort to return to the airport.

Post-flight analysis

The captain filed an ASRS report of his own, which differed somewhat from the first officer's. "During rotation at Newark," reports the captain, he "felt the aircraft settle back onto the runway." He and the FO "concluded there was no damage to the aircraft," so they continued to destination. On the ground at Detroit the two discovered a "scuff" on the tailskid. The captain continues that he "used bad judgment" and "didn't write it up, leaving maintenance to write it up." The captain concludes that "we do not know the cause, but in the future [I'd] simply return to the originating airport for a visual inspection."

The company's director of operations investigated and apparently believed the first officer's version of events. The company assured the FO that he "was not at fault" and that he "did everything short of taking the controls to persuade the captain that it was unsafe to continue the flight." The captain "made serious judgment errors" and "refused input from the first officer, violated the company's standard operating procedures, [and] ignored 'Non-Normal Checklist' items." The airline concluded that the captain "is normally a very safe, conscientious pilot who uses SOPs and follows good Crew Resource Management (CRM)." In this case, however, the first officer "failed in all his CRM efforts to get the Pilot-In-Command to return to land for inspection."

Lessons learned

Your airplane has checklists for unusual or "non-normal" situations as well. The Federal Air Regulations, the Aeronautical Information Manual (AIM), your airplane's pilot's operating handbook (POH), and your own experience and training give you guidance on what to do if you're presented with an abnormal event or indication in flight. Review the guidance available to you, practice it as best and as frequently as you can, and unless you have strong evidence that the guidance is wrong when faced with an unusual situation in the air, take the advice of the persons with the knowledge (and without the stress of facing the situation aloft) and experience to advise you by way of checklists and SOPs.

Case 3: When Checklists Don't Help (ASRS Accession Number 393615)

A Saab 340 experienced a hydraulic failure on short final to landing at Boston's Logan International Airport. Because the airplane was already in the landing configuration and was unable to quickly "clean up" for a go-around, the crew completed the "memory items" of the emergency checklist and landed the airplane.

Clear of the runway and taxiing in, the captain referenced and ran the "abnormal procedures" checklist for hydraulic failure while the first officer worked with ground control. The captain was also in radio contact with his company's mechanics to troubleshoot the problem. The checklist was interrupted by a "request from ground control to move the airplane from the taxiway," and the crew turned its attention to getting the twin turboprop to the gate.

At the gate and during shutdown, the Saab's parking brake failed; the airplane "jumped the chocks" and ran into a company maintenance truck. The airplane suffered damage to the left propeller.

Preparation

The crew wisely detected the hydraulic failure while on approach and continued the landing after quickly noting that it could not reconfigure the airplane for a go-around. Safely on the ground and troubleshooting the problem, the crew became distracted and failed to completely diagnose the problem before losing total hydraulic control and damaging the airplane at the gate.

In-flight

Again, the crew quickly detected the abnormality, completed the memory items from the Abnormal Procedures checklist, and safely landed the airplane.

Post-flight analysis

The reporting pilot wrote to ASRS that "the checklists for hydraulic problems for this aircraft are multi-page and multi-part with many 'if' and 'and/or' branches. This causes much confusion when attempting to sort problems via [the] checklist."

Lessons learned

When reviewing aircraft checklists, and especially if you write your own checklists for abnormal procedures, be thorough, but beware of making checklists that are so detailed that they'll be confusing in actual practice. Simplicity is the key: simplicity in standard operating procedures and simplicity in checklists to back them up.

Case 4: Think One Thing, Do Another (ASRS Accession Number 444540)

The captain of a Boeing 737 commanded "gear up" as the passenger jet lifted into the Texas sky. The first officer, acting as "pilot not flying," responded by repeating the words "gear up," signifying that he'd done what the captain expected. Shortly afterward, as the captain wrote to the Aviation Safety Reporting System, "I noticed an unusual noise in the nose tire and looked at the landing gear handle. [I] found it in the 'Down' position. I looked at the first officer and restated 'Landing Gear Up.' He responded again 'Landing Gear Up.' I then looked at the flap lever and realized that on the initial call for the gear, [the first officer] had put the flaps up instead.

"All this happened fairly quickly. I believe [the first officer] finally got the gear up before the flaps had fully retracted, but I am not sure. With the initial [nose gear] noise I had decreased my climb to approximately 15 degrees [nose-up pitch attitude] instead of the normal 20 degrees. No stick shaker or degradation of flight [performance] was noticed."

Preparation

At first, this sounds like a case of first officer unfamiliarity with the airplane. The reporting captain, however, states that the FO is "highly qualified" with "over four years at the company" flying 737s.

Instead, the reporting pilot suggests that stricter checklist discipline and attention to detail could have prevented this out-of-sequence event.

In-flight

The captain called for gear retraction, but the first officer inadvertently retracted the flaps instead. The captain noticed an "unusual noise" coming from the area of the nose landing gear, likely the squeal of high-speed slipstream air around the nose gear—an unfamiliar noise, since the landing gear is normally retracted at climb airspeed.

Noting some discrepancy, the "pilot flying" reduced pitch attitude. This is a normal response to an abnormality, since in many cases, airspeed can become critical in a climb. In fact, this would have decreased the jet's climb rate (helping overcome the effects of inadvertent flap retraction) while increasing the noted discrepancy, landing gear slipstream noise. After what was likely a very short time, the crew discovered what had happened, retracted the landing gear, and continued with an otherwise uneventful flight. Happily, airplane weight and outside environmental conditions were such that the 737 did not suffer a critical loss of climb performance with the inadvertent flap retraction.

Post-flight analysis

Instead, the reporting pilot credits this error to complacency. "This was the third leg of the first day of a four-day trip," the captain writes in self-critique. "Neither of us were tired. There was nothing out of the ordinary that I can tell, other than my first officer did admit later that he had a headache.

"The only other thing we might attribute to this incident," the captain sums up, is the first officer's "very high comfort level" with the airplane and procedure, "which might have lowered his concentration."

Lessons learned

The lesson of this ASRS write-up is complacency. The more familiar you are with an airplane and its operation, the easier it becomes to do things from rote memory. This is good in many ways, as we've seen in other ASRS incidents, because it'll help you when you're stressed, or you're fatigued, or when multiple status requires you to prioritize your actions and perform tasks quickly.

However, when your confidence becomes overconfidence and you stop paying attention to what you're doing, your experience can slide into complacency, and you may only *think* you're doing what you meant to do. Follow your standard operating procedures and double-check your actions by using printed checklists. Don't get so comfortable with flying the airplane that you feel you don't need to verify everything you do.

Case 5: Flying Head Down (ASRS Accession Number 443613)

Weather conditions were marginal VFR with haze and scattered clouds as the commercial jetliner began accepting vectors for an instrument landing system (ILS) approach at Philadelphia. ATC cleared the flight to descend from 6000 to 3000 feet and provided a heading of 300 degrees to intercept the localizer for KPHL's Runway 27 Right.

The flight's first officer, who submitted the ASRS report, states he began slowing the airplane "because of our proximity to the airport," then he "took my attention to a different page of the Flight Management computer to determine our lateral distance to the localizer." He was "surprised to see we were on [the localizer] but with a 300-degree heading. I quickly verified this information with the localizer [needle of the horizontal situation indicator] and began a left turn to intercept."

What was the other pilot doing during this time? "The captain was busy with Air Traffic Control and [the Before Landing] checklist, and did not 'call' the localizer until we had already passed through it.

"Because of the airplane's speed," the first officer continues in his account, "and my late turn, we went through the localizer and were turning back toward the southwest to intercept when we received a Traffic Advisory, followed shortly" by a Traffic Collision Avoidance System II, or TCASII, alarm. "My recollection is that the TCASII first said 'Reduce Vertical Speed' followed shortly by 'Don't Climb,' and I saw the red lights on the top portion of the vertical speed indicator." In TCASII-equipped airplanes, the system scans for other air traffic and then provides pilots with altitude and heading guidance to avoid a collision. TCASII favors altitude avoidance over vectors around traffic.

"ATC gave the other aircraft (I believe a company 737) a turn when they saw the two aircraft getting close. I continued to slow [the airplane], turn, and descend. At about 3500 feet MSL we could see the airport, and ATC cleared us for a visual approach. To my knowledge no other action was taken."

Preparation

The crew of this jetliner appears to have been task-saturated, both pilots so caught up in the minutia of flying that neither was seeing the big picture. Better coordination of flying tasks in the terminal environment and better cross-checking during critical operations like intercepting the localizer for the instrument approach could have prevented this near collision.

In-flight

"I was not able to focus on the TCASII," writes the reporting pilot. Meanwhile, the captain was "head down" in his checklist and working the radios and was not monitoring his first officer's control of the jet. The crew needed to better prioritize its actions, do whatever was necessary for the first officer to read and interpret the glass cockpit displays while acting as "pilot flying," and pace its actions so the "pilot not flying" was freed up to monitor the airliner's flight path. There is a difference between two pilots flying in an airplane and a two-person crew actively managing a flight.

Post-flight analysis

The first officer has no idea "how close we were laterally or vertically" from the other airliner. He credits the pace of flight into a major jet terminal as a factor in this incident: "We were set up high and fast by ATC when [we were] cleared for the approach. If ATC had given us our position [such as one-half mile south of the localizer] when cleared for the approach, it would have alerted us to imminent capture of the localizer, and would have been very useful." The reporting pilot made no more comment about division of workload, the responsibilities of the "pilot not flying," or his own need to speak up if he is unable to read instrument displays or completely handle the "pilot flying" chores. Only the automated TCASII system prevented what could have been a midair collision involving hundreds of passengers and people on the ground.

Lessons learned

Flight in marginal or instrument meteorological conditions, and at any time during arrival at a busy airport or in heavily trafficked airspace, can overwhelm the unprepared pilot. Note in this instance how even two professionals flying a top-of-the-line jetliner were unable to keep up with the action.

Always remember to pace yourself. Do what you can of your Approach and Before Landing checklists while descending from cruise altitude. Delay "head-down" checklist steps when within 1000 feet of leveling off or when intercepting an instrument approach course. Develop a personalized system for marking your place if interrupted in mid-checklist or for going back to the top of the "open" checklist to confirm your actions when time again permits you to do so.

Case 6: Altimeter Setting (ASRS Accession Number 449130)

A Boeing 727 crew was earning its pay. Departing out of Minneapolis/St. Paul, it was deviating left and right of course during the climb, looking for the best ride around thunderstorms that were sprouting up seemingly everywhere.

The flight was cleared to cruise at Flight Level 230, approximately 23,000 feet above sea level. The ride at FL230 was bumpy, however, and the captain asked for a descent to Flight Level 190, where it had been smoother during the climb.

Leveling at what it thought was FL190, the 727 crew was advised by Minneapolis Center controllers that it was 300 feet off altitude. Quickly, the crew discovered it had failed to set the altimeter to the 29.92 inches of mercury that is the standard altimeter setting above 18,000 feet. The crew reset the altimeter and immediately corrected back to FL190.

Preparation

Flight in the flight levels is an everyday occurrence for jet pilots. It's easy for some of the more miniscule items like altimeter setting to slip if the pilots get complacent. It's even more likely when the cockpit crew is very busy, like it had been trying to get the best ride for their passengers in back.

Even more so than under normal circumstances, flying in a high-workload environment demands that pilots follow standard procedures (such as resetting the altimeter when passing through Flight Level 180) and use printed checklists to confirm their actions.

In-flight

When prompted by air traffic control, the crew quickly discovered its oversight and made a correction. There was no traffic conflict.

Post-flight analysis

The flight's captain, who filed the ASRS account, made no effort to critique his crew's actions, instead merely stating matter-of-factly the point that the crew forgot to set the altimeter.

Lessons learned

This type of occurrence happens every day. The point is to develop and maintain good operating practices. If you have an operation that is unique to the type of flying you do (such as flight-level flying in the case of a turbine or turbocharged airplane), look for a checklist that will remind you of the little things that may mean a lot in unusual circumstances. If your airplane's handbook doesn't have a checklist to cover your unique operation, do some research in similar airplane handbooks. Talk to instructors and pilots more experienced in the operation, and put together your own checklist to use on those days when weather or other factors make you subject to the distractions of pilot workload.

Case 7: Nonstandard Deicing Technique (NTSB Report Number SEA00FA039)

A Mooney 20K collided with terrain approximately a mile and a half from the end of Runway 29 at Pendleton,

Oregon. The pilot and his passenger died in the fiery crash, and the Mooney was destroyed.

The pilot had departed Billings, Montana, earlier that day, flying VFR with an intended destination of Creswell, Oregon. En route, conditions worsened; the pilot picked up an instrument clearance and changed his destination to Pendleton.

On approach to Pendleton he advised air traffic control that he was in icing conditions. He also called ahead to ask for after-hours refueling; after a successful approach, he landed, parked on the ramp, and watched as linemen topped the Mooney's fuel tanks.

According to the National Transportation Safety Board (NTSB), "the pilot borrowed a 2-inch × 6-inch wheel chock from the refueler and—according to the refueler—'started pounding the leading edges of the wings and vertical stabilizer'" in an attempt to remove the ice that had built up during his approach. "After finishing the vertical stabilizer," again according to the refueler, as quoted in the NTSB factual report, the pilot "made the comment that his airplane was now deiced."

"The refueler," himself "a highly experienced general aviation commercial pilot, did not specifically observe the pilot knocking the ice off the leading edge of the horizontal stabilizer.

"During [a post-crash] interview by [NTSB] investigators, the refueler stated that about a half inch of 'somewhat moist snow' had accumulated on the wings of the airplane prior to the time he refueled the airplane, necessitating brushing snow away from the fuel caps" to service the airplane. "After wiping the snow off," continues the account, "there was still a covering of rough ice on the upper surface of the wing. During the refueling, [the line serviceman] noticed that none of the ice on the wing had melted." At the time of the accident, the ambient temperature was $-1°C$.

In a written statement the refueler says that "'after refueling, I looked at the right wing leading edge and pointed out some rough ice behind the area that [the pilot] had cleared, both on [the] top and bottom of the wing. Rubbing his hand over the rough area, [the pilot] commented that there wasn't enough ice to make any difference.

"'He then paid for the fuel, requesting a cash receipt. I went to the office to complete a receipt and then returned to the aircraft. I did not leave my vehicle and did not observe whether or not any additional snow or ice had been removed from the wing. The aircraft departed immediately. I observed the aircraft taking off about one-third of the way down the runway. It appeared and sounded normal' during the takeoff."

Instrument meteorological conditions (IMC) prevailed. The Mooney pilot requested and received an IFR clearance to fly at 10,000 feet en route to Eugene, Oregon.

The NTSB report continues: "The Air Traffic Controller observed the airplane using most of the 5581-foot runway during its takeoff roll. After losing sight of the airplane and issuing a handoff to contact departure control, [the controller] heard the pilot radio 'Mayday, mayday, going down.' There were no further communications from the airplane.

"The airplane had been cleared for takeoff approximately two minutes before the pilot declared a 'Mayday.' The airplane struck the ground and burned about 1.5 miles north–northwest of the departure end of the active runway."

Preparation

This is a somewhat different case because it does not involve a checklist per se, but it *does* result from a very serious failure to follow a standard operating procedure.

Federal Air Regulations do not specifically require ice removal in light general aviation airplanes operated for noncommercial purposes. Failure to remove ice has, however, time and again been presented as evidence that the pilot was operating in a "careless and reckless" fashion. Certainly, ice removal is one of the standard topics taught in flight instruction and is a frequent topic of discussion in print discussions of lightplane flying. It's extremely unlikely that the pilot of this Mooney did not know how risky flight with an iced-up airframe can be and that it is standard operating procedure to completely remove airframe ice before taking off. Whacking off chunks of ice with a wooden chock is definitely *not* a "standard" procedure for removing airframe ice, nor is rubbing a glove over "rough" ice on the top and bottom of the wing and pronouncing that there "wasn't enough ice to make a difference."

In-flight

The pilot and his poor passenger never had a chance once the pilot committed them to takeoff. The Mooney was likely quite touchy in pitch and perhaps roll, and the two persons on board suffered over a minute in a likely gyrating airplane with the almost certain knowledge that they would not survive. A proper preflight deicing or a "no-go" decision would have been the only actions that could have prevented this tragedy.

Post-flight analysis

There obviously was a lot more going on that night that compelled the pilot to attempt this last, fatal trip. Was there pressure to make it to destination, for work or social reasons? Did the pilot need to show his bravado to the passenger, so much so that a delay in taking off seemed unthinkable? Had the pilot survived icing conditions before (a common event in the Pacific Northwest) and convinced himself that he could handle anything? Was this pilot's flight education so poor that he really didn't understand the significance of airframe ice?

Case 8: Doors Not Latched for Takeoff (NTSB Report Number FTW00LA069)

A solo pilot escaped injury in the crash of his Cessna 310Q at Spring Branch, Texas. According to the NTSB factual report, the pilot was departing on a VFR flight when the airplane "impacted a ditch and a fence during an aborted takeoff."

Writes the NTSB: "The pilot stated that during the takeoff roll, at 82 knots [indicated airspeed], the cabin door opened. He attempted to close the door but was unsuccessful, and [he] elected to abort the takeoff with approximately 200 yards of runway remaining. The airplane skidded off the departure end of the runway, and the nose gear collapsed as the airplane impacted a ditch. The airplane sustained structural damage to the right wing, left main landing gear, and [the] propellers.

Preparation

Cabin doors popping open create a huge amount of noise and distraction. In low-wing airplanes, they tend to open right at rotation speed, as the low-pressure area created on top of the moving wing grows rapidly and can overcome a less-than-fully-latched door mechanism. According to most airplanes' pilot's operating handbooks, airplane performance is usually affected only slightly with a cabin door open in flight. The standard operating procedure is to continue the takeoff, fly a normal traffic pattern, land, and then properly secure the door on the ground.

The pilot of this airplane didn't follow that sound advice and instead "attempted to close the door" while the airplane was "passing through 82 knots," which is very close to the 310's published rotation speed.

The departure runway was not unusually short, but it was short enough that it presented a challenge for the 310 pilot, and in the end it was not long enough for the manner in which the pilot tried to use it. Before lining up for takeoff, the pilot should have mentally briefed himself about where on takeoff it was acceptable to abort and at what point (physically along the runway and at what indicated airspeed) he was committed to a takeoff so long as both engines were working. At 82 knots, which is near the 310's normal speed on touchdown after landing, it likely would have taken far more than the "200 yards of runway remaining" to come to a stop. The pilot might have the required distance beforehand using the pilot's operating handbook Accelerate/Stop performance chart, or if that wasn't available, by using the Landing Distance chart to calculate runway remaining requirements at rotation/landing speed. Either would have told him that 200 yards was not an acceptable distance for an aborted takeoff for almost any reason other than an engine failure.

In-flight

There was no in-flight phase in this mishap.

Post-accident analysis

Runway remaining questions notwithstanding, the pilot allowed himself to be distracted by the door popping open on takeoff, and instead of continuing to "fly the airplane," he diverted his attention to trying to close the door until it was too late.

Lesson learned

A quick glance at a Before Takeoff checklist might have warned this pilot to check "Doors and Windows— Secured" before advancing the throttles. Don't forget to use your checklists!

Avoiding Miscellaneous Incidents

There are as many ways to forget steps or miscontrol an airplane as there are pilots, switches, and relays in an airplane. Your only defense against missing miscellaneous oversights, any of which might prove disastrous given the wrong set of circumstances, is to have solidly developed and practiced standard operating procedures. Follow and use checklists for those times when your workload or your condition prevents you from remembering everything there is to do to fly safely.

5

The Ultimate Checklist and Procedures Accident: The Gear-Up Landing

Perhaps the ultimate expression of the failure to follow checklists and standard operating procedures (SOPs) is to land a retractable-gear airplane without extending the landing gear. This type of incident is all too common and does not always discriminate between pilots based on their level of experience.

Sometimes it's the complacency of familiarity brought on by hundreds or even thousands of hours in a single type of airplane that leads to a gear-up landing. In other events it's a combination of distraction and fatigue. In yet more it's a bad-habit pattern, never identified and fixed in initial pilot checkout or in flight reviews, that eventually catches up with the hapless pilot.

Gear-up landings rarely cause serious injury, and many times even the damage to the airplane is relatively minor. They are, however, increasingly costly to repair, and often render the pilot virtually "uninsurable," or insurable only at great price, for years to come. Let's look at some gear-up landing accounts from the Aviation Safety Reporting System (ASRS).

Case 1: Traffic Pattern Distraction (ASRS Accession Number 371403)

A Piper Comanche reported turning downwind at an uncontrolled airport as an emergency medical service (EMS) helicopter lifted off from a nearby hospital. The helicopter pilot was monitoring the airport's Common Traffic Advisory Frequency (CTAF) and announced he was transitioning the area and departing to the southwest. The EMS pilot also radioed that he had the traffic on downwind in sight.

The helicopter pilot saw a possible conflict, so he changed his route slightly to pass below and to the rear of the Comanche. He announced this on CTAF also. "About 10 seconds" later, the helicopter pilot radioed that he was clear of the arriving traffic, and he departed the area.

"When I was about five miles southwest of the airport," continues the pilot of the EMS helicopter, "I heard an aircraft report that [the Comanche had] landed gear up on [the] runway. I do not know if my transition near the airport distracted the pilot of the [Comanche] from his checklist or if he had [mechanical] problems. If my presence was a distraction, perhaps an additional check on short final, such as (1) props—as required, (2) gear—recheck down," might have prevented this gear-up landing.

Preparation

Although the reporting pilot in this case was in the helicopter and only speculates that the distraction of his passing led to the gear-up landing, it does bring up the point that the operation of one airplane may sometimes affect the way other pilots operate other airplanes. In the traffic pattern, follow established procedures and maneuver to try to avoid being a distraction. If transitioning an airport, try to fly outside or above the traffic pattern. The EMS mission requires utmost speed and urgency, but even so, EMS pilots can plan beforehand to avoid the traffic pattern if at all possible.

The Comanche pilot, on the other hand, should have reviewed procedures and practiced techniques often enough so that making a final descent to land without extending and rechecking the landing gear would be unthinkable. Pilots tend to revert to rote memory when distracted or stressed, so that rote learning has to be "right," producing the desired result.

In-flight

Both the pilot of the Comanche and the pilot of the EMS helicopter were following good radio procedure, which prevented a conflict in the airport traffic pattern. Stricter adherence to a standard landing gear extension and verification procedure likely would have saved the Comanche pilot a lot of embarrassment and expense.

Post-flight analysis

It's vital that pilots of retractable-gear airplanes concentrate on the most obvious of tasks: extending the landing gear before making contact with the ground. A busy traffic pattern should serve as a warning to be especially careful to follow good gear-extension procedure. The helicopter might have been a distraction on the downwind leg and caused the Comanche pilot to be concerned about collision avoidance and possibly depart the pattern long enough for the emergency helicopter to pass. But it should not have been so great a distraction that the pilot completely forgot to extend the landing gear before touching down.

Case 2: When "By the Book" Doesn't Fix the Problem (ASRS Accession Number 384410)

The commuter turboprop entered a base leg, and the first officer moved the control to extend the landing gear. Quickly, "we noticed our '3 Green' lights did not appear. We cycled the gear [switch] again and the same outcome occurred, no 'gear down' lights. We flew a missed approach and [ATC] vectors while we ran the emergency checklist.

"That's where our problem began. The checklist called for emergency extension of the [landing] gear," by pumping the hydraulic system, "but we 'knew' the gear was down [because] we heard it." Following the emergency checklist, the crew also "verified 'gear down' by moving [the] power levers back to flight idle [one at a time], and no 'gear unsafe' horn was heard. We also set the flaps to 20 degrees and no 'gear unsafe' horn was heard.

"Therefore, we were satisfied that the gear was down and the problem was with our 'gear safe' lights. The tower visually checked our landing gear" while the flight was on final approach, and the gear "appeared [to be] down. No emergency was declared, yet we had all rescue vehicles standing by as we landed." The flight landed without further incident.

"Once on the ground we noticed the circuit breaker for the gear lights was out. A quick and simple fix—that was the problem. Under the pressure of flying the aircraft and following the proper procedures, there is at no time any reference to the [gear lights] circuit breaker. Yes, we should [have known to] check the circuit breaker—a good idea, but not a part of our trained procedures. We ran all the proper checklists and even spoke with our maintenance personnel" before finally landing the turboprop, but "there was never any mention of a popped circuit breaker. My suggestion is to include a checklist for 'No Gear Down Indication'" to prompt pilots to "check the circuit breaker."

Preparation

The crew in this example was obviously very well trained. It knew the written procedure and followed it to the letter. The flaw in this case was in the procedure itself—and in the crew's blind compliance, without investigating other, and in this case simpler, causes of the improper landing gear indication.

In-flight

Again, the crew did everything by the book. It extended the landing gear, detected the abnormal indication, missed the approach, and reached a safe altitude at which to investigate. It ran the emergency checklists to include manipulating the power levers and flaps to trigger a gear warning horn (if the gear had, in fact, not been down), spoke with company maintenance in troubleshooting the problem, and asked for emergency equipment to be standing by just in case. The fault was in the design of the emergency procedure itself and with the company's training, which did not include checking for a popped landing gear indicator circuit breaker.

Post-flight analysis

The reporting pilot suggests a short "No Gear Down Indication" checklist be added to the company's procedures and training.

Lessons learned

You can bet that this crew now knows the simple step of checking the gear indicator circuit breaker in case either pilot ever faces a similar situation again. The challenge is in getting the word out to other pilots. The checklists provided by airplane manufacturers and in this case airlines are an excellent starting point, but they do not necessarily envision every possible situation you might face in the air.

Seek out knowledge about your airplane's systems. Take a course if it's available; study the pilot's operating handbook and other factory documents to learn all you can, just in case you need that information some day. Scrutinize your checklists, especially if they're provided by someone other than the airplane manufacturer, to see if there are any additions you need to make. Don't make your procedures overly complicated, but make them detailed enough that they'll prompt you to make the right checks under the stress of an in-flight emergency.

Case 3: Landing Gear "Oversight" (ASRS Accession Number 391935)

The Cessna Caravan amphibian entered a left downwind for a "ground" landing. "Immediately after referring to my 'Before Landing' checklist," reports the pilot to the ASRS, "I was distracted by a passenger, and [by] ATC giving me further traffic advisories. This caused me to fail to put the gear lever into the 'down' position, after which I proceeded to land on the floats, wheels up."

Preparation

Amphibious airplane pilots face the additional procedural imperative to be "right" in their landing configuration, because landing with the gear "down" in water is even more disastrous than landing with the gear "up" on the ground. It's even more critical, then, for "amphib" pilots to follow checklists, since their practice and experience may not reinforce what's needed to make the next landing a successful one.

This Caravan pilot might also have briefed the passengers beforehand regarding when it is and is not appropriate to interrupt the pilot.

In-flight

Here, too, we have a gear-up accident resulting from distraction on the downwind leg. Busy airspace and, in the case of amphibious airplane pilots, any landing should make the pilot especially alert to the possibility of a gear-up landing. This heightened awareness, hopefully, would prompt pilots to carefully follow procedure and use checklists as a backup.

Post-flight analysis

Traffic pattern distraction is a common theme in airplane accidents, more so in the case of gear-up landings. With a little tact, it's possible to brief passengers before take-off that there are times in the conduct of a flight when the pilot wishes not to be distracted. A nod, a word on the intercom, or a signal of some sort might tell the passengers not to interrupt the pilot until more than 1000 feet above the ground, within 1000 feet of leveling off in either a climb or descent. You can also use these signals when on an instrument approach or when in the airport traffic pattern. Passengers are familiar with similar restrictions (seat belts, electronic devices) when flying on commercial airlines, and they will readily accept your direction if you provide it to them.

From there, it's a matter of good procedural flying, with printed or mnemonic checklists for verification, to prevent cockpit distraction from resulting in an accident.

Lessons learned

Do what you can to minimize cockpit distractions during the high-workload takeoff, initial climb, approach, and landing phases of flight. Brief your passengers on when it is and is not appropriate to ask questions or make comments. Be sure they know that you're *always* open to help if they think safety's an issue, but otherwise they should only approach you with questions or comments when you're in cruise flight.

Case 4: Practice Wasn't Perfect (ASRS Accession Number 393344)

The newly hired pilot of a Cessna 340 pressurized twin spent the day practicing normal and emergency procedures "just to experiment with the equipment and get the 'feel' of the airplane." He'd performed several practice instrument approaches at multiple airports, without a view-limiting "hood" or safety pilot, but visually as part of his self-indoctrination to the aircraft. He approached his home airfield for the last approach of the day—a simulated single-engine instrument landing system (ILS).

"I simulated single-engine operations by retarding the throttle only and advised the tower that I would [make a] full stop" landing, the pilot later writes in the ASRS. "I intercepted the localizer at the outer marker and was distracted at the marker by the task of getting on course and glideslope." The pilot reduced power on the "good" engine to initiate the instrument descent and forgot to extend the Cessna's landing gear "until a pilot [who] was holding short of the runway said 'landing gear'" on the radio. "I initiated an immediate go-around, advised the tower of the go-around, and flew the pattern to complete the flight with a normal landing."

Preparation

It's commendable that the pilot of this Cessna 340 wanted to become more familiar with the airplane and to increase his flying skills. It's a good idea for all pilots to practice procedures regularly and especially to take the time to get familiar with an airplane when new to the type.

The pilot also apparently has a set procedure of extending the landing gear at glideslope intercept and rechecking it on final approach. But he was not yet experienced enough in flying the 340 to have those procedures firmly ingrained into his habit pattern.

In-flight

When prompted by the pilot of the airplane holding short of the arrival runway, the Cessna 340 pilot immediately broke off the approach, presumably using both engines for a normal go-around, and flew a standard visual traffic pattern to an uneventful landing. The pilot exhibited excellent discipline by not trying to resolve the landing gear problem while on short final in the pressurized twin—many gear-up and gear-collapse landings happen when a pilot attempts to extend landing gear too late to fully lock down.

Post-flight analysis

"My major failure," self-critiques the reporting pilot in the ASRS, "was to retard the throttle a bit at the marker to initiate a descent, instead of leaving the throttle [in place] and dropping the landing gear. I allowed myself to become distracted by the change in performance of the aircraft due to a lack of thrust [brought on by the simulated engine out], and by intercepting the localizer at the outer marker instead of outside the outer marker. My distraction got me out of the rhythm of always doing a pre-landing checklist at the marker, and a 'GUMPS' check on short final. This could have been the most expensive lesson of my life, and practically could have ended my flying career."

Lessons learned

It's easy to forget things when you do them again and again and again and fatigue sets in. Be especially careful when practicing multiple approaches or takeoffs and landings—accident reports are filled with cases that happened "on the second approach" or "while doing touch-and-goes." Each takeoff and landing needs the attention you provide to the first.

Case 5: Flying More Than One Airplane (ASRS Accession Number 396693)

The renter pilot of a Cessna Turbo Skylane RG matter-of-factly reports to the Aviation Safety Reporting System that "I landed an airplane with the landing gear up because I forgot to put the gear handle down. I improperly used the checklist, missing the landing gear part."

The reporting pilot delves into design and human factors that affected the flight as well. "The landing gear handle is difficult to see" in the 182RG "because it is hidden by the flight control yoke. That makes the airplane appear [similar to the] fixed gear type" of the Cessna 182.

Preparation

It's common for pilots, especially those new to retractable-gear-type airplanes, to be renters and fly more than one type of airplane. It's even more critical, then, for pilots new to retractable landing gear to spend some time in the cockpit studying the airplane's layout. A thorough checkout with an instructor very familiar with that particular airplane type should reinforce techniques for gear extension and retraction and help the renter pilot develop habit patterns that will make setting up for a gear-up landing completely foreign to the way the pilot flies the airplane.

In-flight

In the air and preparing to land, the reporting pilot apparently followed the instincts he picked up in fixed-gear airplanes—lower the flaps and make a big power reduction to initiate descent from pattern altitude. A more correct technique in retractable-gear airplanes is to initiate the descent with gear extension and only then steepen the descent with a small power reduction as necessary to maintain desired glidepath. The reporting pilot was used to flying fixed-gear airplanes, where power reduction is the way to generate a descent, and combined with his failure to use a checklist to back up memory items, he reportedly did not even notice his omission until the airplane's belly made contact with the runway.

Post-flight analysis

The reporting pilot recommends that "all retractable [gear] types of airplanes should have at least a visual warning [of unsafe gear position at a landing power setting] on the top part of the [instrument] panel in conjunction with the audible warning." He writes that "I will use a proper checklist for each different type of airplane" he flies "to prevent missing items on the checklist."

Lessons learned

When you transition into a retractable-gear airplane, or if you have experience with one type of retractable-gear airplane and you're transitioning to another, you should seek out an instructor well experienced with that particular airplane type. Insist on a thorough indoctrination, not just the "three times around the patch" sort of RG checkout that unfortunately so many of us get. Especially if you're new to retractable-gear airplanes, if you go from one type of RG to another, or if you vary between flying fixed- and retractable-gear models, the habit patterns you've developed may not always be correct for the airplane you're flying today.

Case 6: Distraction on Downwind (ASRS Accession Number 400169)

The pilot of a light retractable-gear airplane entered downwind for Sky Acres Airport, when a Cessna 150 departing from the same airport reported deer on the first quarter of the active runway. The pilot continued on downwind, making adjustments to land "long," beyond the point where the deer had been sighted. "Thinking I had deployed/lowered the landing gear," the pilot wrote to ASRS, "I extended my approach to a point beyond the deer sighting and landed." Unfortunately, in changing his normal procedure to land long, the pilot forgot to extend the landing gear.

Preparation

The pilot wisely decided to change his aim point on landing to avoid the likelihood of hitting a deer. Once that decision was made, however, the distractions caused the pilot to deviate from his standard procedures, and he apparently reverted to earlier habits, which did not include gear extension and verification before landing.

In-flight

In this case also, the distracted pilot never noticed his failure to properly configure the airplane for landing until the airplane made gear-up contact with the runway.

Post-flight analysis

As the reporting pilot succinctly puts it, "Cause: Pilot Error. Distraction from the landing checklist."

Lessons learned

Distractions can be very powerful, and they demand a great deal of your attention. When you're distracted you also tend to fall back to rote learning and subconscious action. That's why you must be certain that your basic habit patterns are conducive to properly flying the airplane. This comes only from a thorough indoctrination to the airplane, with sufficient practice to make the "right way" the only way you'll fly the airplane. If the distracted pilot in this case had proper habits, he would not have attempted a descent without the landing gear. Think of how fast he'd be flying in a normal final approach rate of descent with the gear up, or how much more than normal he'd have to reduce power for a normal rate of descent, with the landing gear up. Those discrepancies alone should have warned the pilot that something was wrong and triggered the need to go around and try the approach again.

Case 7: Don't Depend on Warnings (ASRS Accession Number 401885)

An instructor completed a "short instructional" flight with a rental pilot. The newly minted RG pilot was enthusiastic about flying the Piper Arrow; in fact, he asked the instructor to take a picture of him and his friend, who would be his first passenger in a retractable-gear airplane. After the photograph and a brief discussion of fuel loads, turbulence, and winds, the pilot preflighted the airplane, ran the normal checklists, and launched for a short sightseeing flight.

Returning to the airport, as the pilot reports to the ASRS, "I called the tower approximately seven miles [out] and was told to enter a right downwind for Runway 5. There was a Stinson returning from [a nearby airport] to enter a left downwind for Runway 5, as well as an Archer just outside the Final Approach Fix for the Localizer-Back Course [approach to] Runway 5.

"I did a GUMPS [gas, undercarriage, mixture, propeller, switches] check. I was watching for traffic on left downwind and also on the extended final. I already had one notch of flaps [extended] but knew we would be higher than normal, so I pulled the throttle back to 12 or 13 inches manifold pressure, when I would normally use 17 inches. I put in the second notch of flaps and turned final, and did another GUMPS check verifying mixture to full rich and putting the prop lever fully forward. I do not recall verifying three green [landing gear indicator lights].

"As we were on short final with full flaps [extended], airspeed [was] at 90 miles per hour and dropping, the Archer behind us was cleared to land and the Stinson was told it would be number three to land. I flared and held the flare position, allowing airspeed to dissipate. At the last minute there was a very slight 'balloon' which I was counteracting when we touched down to the sound of metal scraping. As soon as I realized what was happening, I pulled the mixture [control] to lean, turned off the fuel pump, and turned off the master switch.

"As the aircraft came to a stop, the passenger opened the door and said, 'Where was the gear warning horn? I never heard the gear warning horn.' I do not remember verifying 'three green' during the GUMPS [check], but I do remember doing the GUMPS. Even though I pulled power early and farther back than normal, there had been no gear warning horn. At no time while on final [approach] did the gear warning horn sound. No one in the [control] tower gave a warning."

Preparation

The reporting pilot was new to retractable-gear airplanes. He was enthusiastic to the point of wanting a photographic record of his first flight with a passenger and had apparently been schooled in normal procedures and checklist use based on his numerous references to the use of printed and memory checklists.

There is a difference, however, between using a checklist and merely rattling off the steps without confirming critical flight items have been done. "I do not remember verifying 'three green' during the GUMPS [check], but I do remember doing the GUMPS," writes the pilot in his ASRS account. He had been drilled to say the word GUMPS at various points during a landing approach, but he had not really be taught what GUMPS was supposed to prompt him to do.

More concentration on technique during a solid pilot checkout, instead of a "short instructional" checkout flight, might have instilled habit patterns that prevented a gear-up landing when the pilot was in charge and distracted.

In-flight

Given that the passenger got out of the airplane and said "Where was the gear warning horn? I never heard the gear warning horn," it seems that the passenger, too, was familiar with retractable-gear airplanes and proper landing gear indications. The passenger was not the pilot-in-command (PIC), but might have been more attentive to the gear position, knowing that the PIC was new to retractable-gear airplanes. In the best of worlds, the pilot would have briefed the passenger beforehand with something like, "since I'm new to flying this airplane, I'd appreciate any advice or help in watching for traffic and configuring the airplane. Don't hesitate to speak up if you have any suggestions." This could open up the door for full use of the gear-knowledgeable person in the passenger seat, who (if told it was okay to lend guidance) might have been more comfortable about watching for and mentioning any problems with landing-gear configuration.

Post-flight analysis

From this ASRS write-up, it sounds like the pilot was overly dependent on warning systems to remind him to extend the landing gear. Variations of "I never heard the warning horn" appear several times in this account, and the reporting pilot ends it with a statement that suggests he expected the tower controller to protect him from a gear-up landing as well. The pilot also notes that he "pulled the throttle back to 12 or 13 inches manifold pressure when [he] would normally use 17 inches," and that the airplane "ballooned" in the flare, but abnormal sensations point to a less-than-normal drag configuration in the landing. Not resolving these discrepancies contributed to the gear-up landing as well.

Lessons learned

Warning horns, lights, and outside observers all serve as an emergency backup should you forget to extend the landing gear. Nothing, however, can replace good technique, including the practice of when it is appropriate to extend the landing gear and looking for the signs (unusually high airspeed or the need to reduce power more than normal) on approach as a means of confirming the gear is down. And, of course, all the GUMPS checks in the world do no good if you don't actually *check* the gas, undercarriage, mixture, propeller, and switches positions when you call out GUMPS.

Case 8: Electrical Failure Causes Distraction (ASRS Accession Number 409623)

The pilot of a Mooney M20J noticed that a "Low Voltage" light illuminated while on an IFR flight with his wife and two children. The light reset itself and remained extinguished for a while, but then it reilluminated momentarily. The abnormal electrical indications were "interspersed with long periods of normal charging indications." Nonetheless, the pilot elected to land at a large airport to have the charging system inspected by a mechanic.

FBO technicians looked at the Mooney and determined that "the alternator was charging. It seemed, however, to have low output when it was heavily loaded." The pilot and mechanics discussed alternatives, and the mechanic "stated that I could safely proceed to my destination so long as I kept the [electrical] load minimal" and flew under visual flight rules.

After a quick call to the rental airplane's owner to discuss options, the pilot loaded his passengers into the Mooney and proceeded as the mechanic had suggested. The pilot did change his intended destination to an airport with repair facilities "so that the alternator could be further evaluated and replaced if that turned out to be necessary."

After engine start the pilot "noted the alternator was charging" and "as suggested, I kept the electrical load to a minimum—just the transponder, navigation lights, and one radio. I even pulled the circuit breakers on the autopilot and the electric trim." The pilot decided the risk was acceptable. The weather was confirmed to be VFR for the entire route, and "I always travel with a hand-held radio and portable GPS with two extra sets of batteries." He states he felt that "even in the unlikely event there was an electrical failure, I had appropriate backup systems.

"The flight proceeded uneventfully with the exception of difficulty with radio communications in the vicinity of Norfolk [Virginia] and Patuxent Approach having difficulty with our transponder. Subsequent controllers did not have difficulty either with our radios or [our] transponder."

Meanwhile, "the alternator indicator switched to show a continuous discharge indication at some time near our destination. I cycled and then pulled the alternator field circuit breaker as appropriate" to the Abnormal Indications checklist. "I weighed my options and, given our position and the knowledge that we were planning on having the symptoms evaluated" at the destination airport, "I judged that it was best to continue to land" there.

The pilot approached his destination. "Since we were unable to raise the FBO on UNICOM, we chose Runway 30 as our landing runway after receiving the ATIS [automatic terminal information service] from a nearby airport. "During our first landing approach, the approach was too fast to allow for a safe landing, and after a brief touchdown" he initiated a go-around. "This included raising the [landing] gear."

On the second approach, "at the mid-field [point of the] downwind leg when I dropped the gear, a loud screeching noise began. It sounded like metal being torn apart, and my first thought was that something had impacted with the airframe, such as a bird." Another thought was that "something in the engine was binding up." But "when the [instrument] panel lights began flashing, it became apparent that this sound was associated with the battery failing and the [electrical] equipment going off.

"Given the nature of this metallic sound, I was concerned that beyond the electrical failure, I might be dealing with an airframe or engine problem. The radios were dead, and throughout the [traffic] pattern the flaps would not extend beyond about ten degrees." The "previous landing attempt was complicated by a high airspeed on approach," so the pilot's "attention was focused on the flap problem and controlling the airspeed. The first sign of any further difficulty was when the airplane contacted the runway in a gear-up configuration. The airplane slid down the runway, crossed the end of the runway and came to rest in the grass beyond." Luckily, there were no injuries.

Preparation

The pilot delivering this ASRS account had an apparently typical general aviation education prior to this event. In review he writes, "I have thought about how 'emergency training' is structured and I would suggest that more attention be paid to 'compound emergencies.' All the practice I have done on 'electrical failures' has amounted to a CFII announcing that I had a failure in flight somewhere between points A and B. Then I would run through the [electrical failure] drill in an orderly manner, proceed to an airport and demonstrate that I knew how to pump down the [landing] gear and land without flaps. It was all very simple, direct and straightforward, and in retrospect, nothing like the real event.

"Emergencies practiced in the [airport traffic] pattern are usually 'engine failures' on takeoff or landing. I would expand the pattern emergency training to include electrical [and potentially other] emergencies as well, so that one has to really stop what one is doing, consider the situation and then resume a checklist. This would be a more difficult exercise than simply having an 'emergency' checklist to go through while one is simply navigating the airplane and not already in the middle of another checklist.

"Finally," concludes the reporting pilot, "I do participate in a safety program and have found that most of my review training 'emergencies' deal with partial panel instrument work—I will be specifically requesting my instructors throw a more varied list of emergencies at me in the future."

In-flight

The pilot of this Mooney allowed the noise and systems outages caused by the electrical failure to distract him from checklists and procedures to the point that he had to go around once he'd touched down on the first attempted landing. After the go-around, he retracted the landing gear, which put an additional draw on the Mooney's feeble electrical system and led to complete battery failure when he attempted to re-extend the gear on the second approach. All these failures, and the distraction created at a critical point in the flight, simply demanded so much of the pilot's attention that he lost sight of the big picture and instead focused on a small part of the problem. Unfortunately, his attention was not on properly configuring the airplane for landing.

Post-flight analysis

The reporting pilot continues his self-critique: "Several factors combined to lead to this incident. First was my decision to proceed with flight with an intermittent problem. In the past, when on a similar pleasure trip, I had a similar but continuous 'low voltage' flashing indicator. I diverted to an airport and followed a mechanic's advice when [he told me] the aircraft was not safe to fly until repaired. The ensuing six hour delay [on that trip] was nothing compared to the grief and stress this [gear-up landing] affair has caused. Although I felt I could rely on the mechanic to give me sound advice, I will be more cautious in the future with any potential problems.

"Secondly, I learned how difficult it is to actually use a hand-held portable radio in the cockpit. Although I knew how it worked and had tried it out on the ground, I had never attempted using it in the cockpit. During the brief radio difficulties around Norfolk, I tried it and could barely hear it—I needed an earplug (which I didn't have) to be able to hear it over the cabin noise. This turned out to be an issue later, because after my electrical system failed during my go-around, individuals on the ground saw my gear up and attempted to radio me. Although I kept calling positions in the pattern on my dead radio, I obviously didn't hear" the people on the ground. "If I had been carrying an earplug for the hand-held radio, I would have used that instead and might have heard" the callers, "averting the incident.

"The most obvious factor here, which may be the hardest to correct, is how to handle the interruption of a checklist. I had started through 'GUMPS' and got distracted by the noise/electrical failure right after 'Undercarriage.' I never confirmed the 'gear down' part of the checklist. I even have a personal one-step check I 'always' do immediately after turning final, that consists of saying aloud 'On final, gear down,' and checking the status of the gear. I was just so focused on the airspeed and making sure it kept down with the flaps minimally extended that even this [final approach check] was missed.

"Although there was an unexpected distraction right at gear extension, once I had realized [that] the airplane was still flying and the engine was still running smoothly, I should have taken a step or two back and re-run through the checklist. Having been interrupted in that process in the manner I was, there was no way I could have been sure just where I had left off when I was distracted."

Lessons learned

Standard operating procedures can include prethinking scenarios which will automatically ground an airplane. In most cases, an intermittent or unresolved electrical problem should preclude all flight until the problem is fixed.

Once airborne, it's vital that you follow standard procedures for gear extension so you won't suffer the same humiliation and expense that so many, including this Mooney pilot, have felt.

This pilot's experience reminds us that we should go beyond the basic emergency procedures training prevalent in general aviation training. The failures normally presented are done to satisfy the requirements of certificate and ratings checkrides, and not to practice the real-world situations that can lead to gear-up landings and far more dire consequences. It's really up to you to make sure you have the education you need, because "the system" won't totally prepare you for scenarios like this Mooney pilot faced.

Case 9: End of a Long Duty Day (ASRS Accession Number 410913)

A Piper Navajo Chieftain was completing a tourist sightseeing flight from Fairbanks to Fort Yukon, Alaska, when it landed gear-up with no injuries. The pilot had flown 10 of the 11 previous evenings, logging over 49 hours in 11 days' time, and he wrote to the ASRS citing fatigue as "the major contributing factor" in this incident.

"The VMC [visual meteorological conditions] flight," reports the Navajo pilot, "was normal until the arrival at Fort Yukon." The pilot extended flaps "normally, at the south shore of the Yukon River," and reduced power "to allow the airspeed to stabilize around 130 knots indicated airspeed for landing gear extension. At this point, either a comment about the village or a passenger's question distracted me," and although the pilot reports he was using a checklist, he did not put the landing gear handle to the "down" position. "The approach was normal until short final when a flock of large birds took flight from the area of the [runway] threshold lights.

"Birds are fairly normal at this location," continues the reporting pilot, "and are normally a 'non-event' as they disperse. This time, however, two of the birds flew down the runway toward the touchdown area. As we were overtaking them at a high rate of speed, I began to sit very high in the seat to monitor their location. This action was occurring at the same time as the final reduction of power and normally the final gear check before landing.

"The aircraft is equipped with a gear warning horn but it is not connected to the headset audio circuit. The warning horn is barely audible when the engines are running and a headset is worn."

Preparation

Again, distraction and failure to follow established procedures are the ultimate contributors to this incident. The pilot's fatigue made it all the easier for him to succumb to the effects of these distractions. With the prior knowledge that the airplane's landing gear warning horn was inaudible with the engines running and with a headset on, the pilot should have been more cautious about following procedure and double-checking landing gear position before touchdown.

In-flight

The reporting pilot also might have better briefed passengers about avoiding distractions at critical points in the flight and in general have done more to comply with the "sterile cockpit rule" required under the governing FAR Part 135. This, and more emphasis on basic procedures and cross-checks, might have prevented this gear-up landing.

Post-flight analysis

The pilot critiques his performance: "Do I regularly use the checklist? Always! I learned to fly in a military flying club and have taught in several military flying clubs. I used a checklist as a crew member in the military, and as a test engineer for a manufacturer. Closer adherence to the 'sterile cockpit rule' (FAR 135.100) and checklist usage (FAR 135.83) may help to alleviate the situation. But requiring aural warning systems (landing gear warning, stall warning horns, etc.) to be part of the audio circuit [headset] would provide timely warning of an unsafe condition."

Lessons learned

There's no substitute for checklist use and adherence to standard operating procedures to avoid landing without benefit of the airplane's landing gear. This flight flew under FAR Part 135, more stringent than the Part 91 rules governing private flight, and Part 135 includes specific guidance for pilots aimed at improving checklist use and reducing cockpit distractions. Private pilots would do well to review Part 135 requirements and consider voluntarily adopting their provisions, as a means of further reducing in-flight risk and improving their professionalism even if they are not being paid to fly.

Case 10: Turbulence Changes Technique (ASRS Accession Number 415241)

A Beech A36 Bonanza pilot was approaching destination in heavy turbulence. As he wrote to the ASRS, "I was distracted during my landing checklist. Turbulence caused me to pull back in the power early. This slowed the airplane into the flap operating range.

"Usually," he continues, "you have to slow the airplane to flap operating speed by lowering the landing gear [to create aerodynamic drag]. I did not notice the wheels were not down and locked until I landed. I was the only one on board and was not injured."

Preparation

Knowing that the turbulence would require a change in his standard operating procedure, this pilot might have been able to anticipate the need to double-check his landing gear before landing. Better before-landing technique, which should include verification of safe landing gear indications, also could have helped avert this mishap.

In-flight

Like almost every retractable landing gear design, the A36 Bonanza has audible and visual warnings of unsafe landing gear configuration. Full flap extension without the gear down, or in some models reduction of the throttle to near-idle, causes "gear unsafe" warnings to sound and light in the cockpit. Greater gear vigilance should have helped the pilot of this airplane avoid a dangerous and costly oversight.

Post-flight analysis

The pilot did not offer any self-evaluation or critique in the report to the Aviation Safety Reporting System. He might agree, though, that better-practiced procedures and stricter adherence to established prelanding check-lists are the answer to avoiding a gear-up landing.

Lessons learned

No flight is perfect. Every airplane operation faces something out of the ordinary. If you have firmly established procedures and good checklist discipline, you'll be able to recognize when you're doing something out of the ordinary. Use that knowledge as a cue to be even more careful to check and recheck critical flight items.

Case 11: Long Day, Short Rollout (ASRS Accession Number 418627)

An amphibious Cessna Caravan turboprop settled down the final approach course of an ILS approach. The Caravan's pilot, on "the fifth flight of a twelve-hour duty day that included four earlier approaches to minimums at different, unfamiliar airports," flew the big single down to decision height, only to have to miss the approach because the runway was not in view.

"At the missed approach point," the reporting pilot wrote to the ASRS, "I had ground contact but could not see the runway environment, so I executed the missed approach procedure. I retracted the landing gear and began to climb. I advised [air traffic control] of my situation" and requested he be allowed to enter the published holding pattern. "After talking to someone on the ground" at the intended destination airport, he told ATC, "I would possibly try [the approach] again."

From the holding pattern the reporting pilot tried to reach someone on the UNICOM frequency. No one was in the airport office, but the pilot did reach the pilot of another company aircraft, which was on the ground awaiting IFR release once the reporting pilot's Caravan cleared the airspace.

The grounded company pilot advised that "for some reason, the runway lights were at low intensity, and that [was] probably why I had to miss. They said the lights were once again on 'high.' Perhaps I had turned them down inadvertently during the [first] approach when I clicked the [transmitter microphone] seven times to turn [the runway lights] on." The company pilot on the ground told the Caravan pilot "they would monitor the lights and ensure that they were at full intensity if I wanted to make another attempt.

"I told Center that I would like to try the approach again. Once again I was cleared for the approach and began to fly it as published. Once established inbound, I became concerned about the runway light status and made several attempts to contact the aircraft on the ground, but they did not answer. About three-quarters of a mile from the missed approach point, I had the runway in sight and began my [final] descent for landing. I would soon learn that, during the second approach, I failed to lower the landing gear at the final approach point." The Caravan slid onto the pavement and into the grass, without injury or serious damage.

Preparation

Multiple approaches and repeated attempts at the same instrument approach appear frequently in accident and incident reports. When a pilot misses an approach only to try the same approach again, there's a natural temptation to try to "go just a little lower" or "fly just a little farther" to try to make a successful landing.

As a rule of thumb, it's extremely risky to fly a second attempt at an approach unless the pilot either (1) flew sloppily on the first attempt and has reason to believe he or she will do better next time, or (2) has reason to believe that the environmental conditions of the approach will significantly improve the second time around. The pilot of this Caravan arguably did have good reason to try again—the runway lights weren't turned up on his first approach—and, in fact, his gamble worked, because he was able to see the airport and complete his instrument approach.

Unfortunately, the pilot followed the ILS to a point where he could land visually, but he didn't follow his normal technique in doing so. He forgot to extend the landing gear. Simply knowing that he was flying an inherently risky procedure (the repeated approach) should have cued the pilot that he needed to very carefully follow procedures and use checklists to verify his actions.

In-flight

During his second approach, the pilot admits he was distracted, "concerned about the runway light status" and making "several attempts to contact the aircraft on the ground." He was not focused completely on properly flying the airplane. More concentration on the mechanics of flying the Caravan and following the ILS might have helped this pilot avoid the gear-up landing.

Post-flight analysis

Not only was the pilot dividing his attention between the task of flying and his attempts to radio the ground, he was also likely fatigued. He'd flown four tight approaches over the previous 12 hours and was flying his sixth approach when he landed gear up.

"Several factors," the reporting pilot critiques, "including my preoccupation with the runway light status, contributed to this incident." He also "failed to reset the electronic 'Heads Up' checklist to the 'Before Landing Checklist' section during the missed approach, and instead reverted to a mental checklist that failed to identify the proper [landing] configuration. In the future, I will take the time to reset the checklist to the current [correct] phase of flight or at least revert to the 'paper' checklist to…never fly past the final approach point without ensuring the aircraft is correctly configured for landing."

The changing nature of the airplane's amphibious gear was also a factor. "The Caravan is normally a fixed-gear airplane and we fly it in that configuration seven months out of the year. Strict attention to the checklist is necessary to avoid this mistake." Further, "I failed to heed the advice of the aircraft's gear advisory system. It makes an automatic announcement that either the 'gear is up for water landing' or 'gear is down for runway landing' during the approach. I was aware that the advisory was sounding off but I failed to differentiate between the 'water' versus 'runway' landing announcements.

"On top of everything else, fatigue certainly affected my performance during this flight. It was an unusually busy flying day for me. I was not under any pressure to complete this trip and I felt reasonably alert and competent to do so. However, the long day, coupled with the increased workload of several IFR approaches to multiple unfamiliar airports at night and as a single pilot, took its toll that [culminated with] this incident."

Lessons learned

Multiple approaches, low IFR weather, nighttime conditions, and pilot fatigue are all warning signs. If you elect
to attempt a flight under any of these circumstances, do
so knowing that you are wholly dependent on your
long-practiced flying skills for a safe outcome. Your only
defense against these hazards is to religiously use
checklists as a reminder to do everything you need to
do. Cockpit automation and warnings, like the Heads
Up checklist and the automated landing gear position
announcements in this Caravan, only work if you set
them up correctly and you listen for and heed their
advice.

Avoiding Gear-Up Accidents

Landing without extending the landing gear before
landing a retractable-gear airplane is probably the ultimate expression of the failure to use checklists/failure
to follow standard operating procedures mishap. There
are almost as many different ways to end up landing
without benefit of landing gear as there are pilots who
fly retractable-gear aircraft. The only way to avoid landing gear-up is to have good and well-practiced piloting
techniques and to use mnemonic, printed, or automated
checklists to make sure the gear is locked down before
the airplane meets the runway.

Conclusion: Avoiding Checklist- and Compliance-Related Mishaps

Printed and mnemonic checklists and procedures are a great help to us when we fly any airplane, whether it is a small single-engine trainer or a multipilot jet airliner. There's a lot to remember when flying in busy airspace and/or in challenging weather, and checklists can keep us from missing items crucial to the safe operation of the airplane.

Checklists really have several different functions, which add up to overall flight safety. Checklists

1. Present actions in a logical order

2. Set the pace for pilot actions, by directing critical items first and "clean up" items to follow

3. Provide a quick plan of action for emergencies

4. Provide for "quality control" in the cockpit, if used to verify no action was overlooked

5. Give the pilot the benefit of the airplane designers', engineers', test pilots', and other pilots' experiences with the make and model of airplane, by including

> extra steps discovered to be important during the
> production and operation of the airplane

Good checklist use is essential to a safe flight.

Danger Signs

Looking at the Aviation Safety Reporting System write-ups and National Transportation Safety Board preliminary reports cited earlier in this book, it's possible to come up with a list of "danger signs." These are clues that a pilot might detect that show that a checklist- or standard-operating-procedure-related mishap is a possibility. Seeing a danger sign, you might recognize that you need to change the way you're doing things to avoid a level of unnecessary risk. What are some of the common danger signs?

Fatigue

One of the more frequently noted contributing factors in checklist- and compliance-related incidents is pilot fatigue. We all know that we're less capable and less sharp when we're tired. The difficulty is in recognizing fatigue in ourselves and judging when we've had enough and need to rest before flying. Remember, too, that you need to project the effects of fatigue into the future—you may feel rested enough to take off now, but will you still be effective flying a tight instrument approach or into a dark, mountain airport after two or three hours of flying the airplane? Checklists are your best defense if you find yourself getting tired in the cockpit, because they'll remind you to do things that you may have missed in your fatigue-degraded performance.

Complacency

Likewise, another common contributing factor is pilot complacency, or lack of attention to flying brought on by familiarity with the task. Checklists and standard operating procedures (SOPs), or simply doing things the same way every time, should help you stay focused and avoid the performance-robbing effects of overfamiliarity and overconfidence.

Distraction

Almost all checklist-related incidents result from some level of pilot distraction. Brief your passengers beforehand on when it is and is not permissible to interrupt you with questions or comments. Pay attention to what's going on around the airplane, but don't let outside distractions draw your attention away from your responsibilities as pilot-in-command (PIC). If you're distracted away from a checklist in midstream, return to the beginning of that checklist and reconfirm that you've done everything you think you have.

Feeling rushed

Quick turns. Weather pressures. Passenger expectations. "Get-home-itis," the "disease" that makes us take unnecessary risk to get where we want to go in a hurry. These can all heighten the stress level in the cockpit, and with increased stress comes the greater likelihood of improper, incomplete, or forgotten actions. Calm down to the best of your ability. Ask yourself if you're making a go/no-go decision based on sound flight planning, or if it's due more to the emotional need to try to get to your destination on schedule. Once you're in the airplane, remember you can't compress time; it'll take as long as it takes, so perform actions correctly and you'll get there as quickly and as safely as possible.

Unusual operations

By definition, an unusual flight operation is one when your practiced standard operating procedures may not work. Flying with inoperative equipment, flying an unfamiliar airplane, or flying into new-to-you territory are all examples of unusual circumstances that have been involved in aviation incidents and mishaps. If you find yourself contemplating an unusual flight, take what steps you can to change your plans to make the flight more normal for you. If you're not able to completely eliminate the unusual characteristics, then approach the flight very cautiously, only in the best weather or under the most controlled circumstances. Review the differences between this flight and your normal procedures beforehand to develop a new standard for the way you'll fly, and be especially careful to review checklists in flight to make sure that the unusual circumstances have not caused you to miss doing something vital.

Multiple or repeated attempts

You miss an instrument approach and decide to try it again. It's your second or third flight of the day, and you elect to skip some checklist steps, either consciously or not. It seems that multiple or repeated actions encourage pilots to miss items, "fly a little lower," or "go a little longer" to try to complete a flight. Be especially careful to follow standard operating procedures and double-check your actions with checklist use if you find yourself making multiple takeoffs or approaches.

Unresolved discrepancies

If things aren't going as you'd expect and you don't discover the difference, you have an unresolved discrepancy. If you have noticed a problem or a difference between flight expectations and actual results, but you do nothing with the information, you have an unresolved discrepancy. Checklists and standard operating procedures will help you check actual performance against projections and give you a precise plan of action for dealing with discrepancies so that they are no longer unresolved.

These danger signs warn you that you need to better focus on procedure and that you should pull out the checklist as time permits to make certain you have not missed accomplishing any critical flight items. Here are some specific techniques you might consider adding to your bag of tricks—additional standard operating procedures that should give you greater awareness of the airplane and its operation and provide you the time and discipline to run checklists and confirm you've done what you think you have.

Suggested Techniques

Here are some suggested techniques you can incorporate to help avoid checklist- and compliance-related mishaps. Feel free to use what you like and modify or throw away the rest—do what's comfortable for *you*, as long as it's safe and it gets the job done.

"Checklists" versus "do lists"

Most of us learned to use checklists by reading a step, doing a step, reading another step, doing that step, and so on. This is a good way to learn the sequence of a large number of unfamiliar steps, but it is not conducive to operating an airplane, especially a very complex one, in busy airspace or in adverse weather.

Consequently, most pilots learn to fly using checklists, but discard them as unwieldy once they pass the checkride, if they make use of checklists even that long. "Do lists" (read a step, then do it) don't accomplish their intended purpose when we don't use them for all phases of flight. Far better is to use them as true "checklists"—accomplish the steps you know it takes to configure the airplane for the next phase of flight, then as time permits pull out the checklist and confirm you have not forgotten anything. You can use printed checklists (as you should for all but the shortest and most time-critical piloting actions) or mnemonics (like the GUMPS check on short final), but use *something* to verify your actions.

Establish, trim, check

As a standard operating procedure that encourages checklist use, get in the habit of "establish, trim, check." Establish the desired flight configuration, trim the airplane for hands-off flight (or use the autopilot, as applicable), and check to make sure you're established as you wish by referencing the appropriate checklist. For instance, after rotation, establish the pitch attitude and power setting necessary for climb, trim the airplane for hands-off climbing at the correct attitude, and then reference the checklist. You'll usually not have to do anything by the time you run through the checklist, but you'd also be amazed at how often the checklist *does* prompt you to turn off the landing lights, or turn on the transponder, or sometimes even retract landing gear or flaps.

Division of attention

Be sure, though, that you don't become so locked in to using a checklist that you ignore what's going on outside the airplane while you do so. Your first responsibility is still to fly the airplane, and a big part of that responsibility is terrain avoidance and to see and avoid other airplanes. Get familiar enough with your checklists that you can check an item or two, quickly scan outside the airplane, then check the next couple of items, and so forth. If you're flying in instrument meteorological conditions (IMC), you'll need to substitute scans of the flight instruments for looks outside the airplane. Remember also to keep the engine gauges and other systems indicators in your scan as well. If you know your checklists and have already done most or all of what the checklist prompts you to do, then with a little practice you'll be able to use checklists in short bursts like this.

Restart interrupted checklists

If a passenger or some outside object distracts you while you're using a checklist on the ground or during one of your scans in flight, mark your place in the checklist if at all possible. If you can't positively remember where you left off, start the checklist over from the beginning when the distraction permits you to return. You don't need to physically repeat actions you're *sure* you've done, but use the checklist to be certain you haven't missed something critical along the way.

Checklist use during taxi

Avoid running checklists while taxiing in single-pilot airplanes. Ground operations are tricky enough, especially at night, in low visibility, or at busy or unfamiliar airports. Keep your eyes and attention outside until you can stop at the run-up area. Then, make your before takeoff checks and prepare to launch.

Many pilots also inadvertently retract the landing gear or make other mistakes by trying to complete checklist items during the landing roll or while taxiing to the ramp. Clear the runway, come to a stop, then take a short moment to "clean up" the airplane and reference the After Landing checklist.

Sterile cockpit rule

To help avoid distractions during critical operations or phases of flight, establish the airline-style "sterile cockpit rule" as one of your standard operating procedures. When the sterile cockpit rule is in force, limit as much as possible any conversation or actions not immediately necessary to establish the airplane in the next phase of flight. Brief your passengers beforehand that there will be parts of the flight when you'll be busy, and you request they do not interrupt you during those times. To make it easy for them, you can explain that unnecessary conversation is restricted once the engine(s) start until you level off in cruise, and then once beginning the descent until the engine(s) stop on the ground. You, as pilot, can make exceptions during long climbs and descents as you wish, but ask your passengers not to take it personally if they speak and you ask them to please hold their comments. Tell them also that you *always* welcome their input if they notice something they think affects the safety of flight, such as seeing another airplane nearby.

Altitude-critical areas

So when do you invoke the "sterile cockpit rule"? When you're in an "altitude-critical area" (ACA). An altitude-critical area is any place within 1000 feet of the ground, or within 1000 feet of leveling off from either a climb or a descent. Start-up, taxi, takeoff, and initial climb all take place in an ACA, so restrict unnecessary conversation or actions until more than 1000 feet above ground level. This is no place to be personally dealing with "nuisance" status like a failed transponder or a sick passenger. They can wait until you're further away from the ground. Similarly, the last thousand feet before leveling into cruise flight will absorb most of your attention, so delay actions not immediately necessary in that last thousand feet, and don't permit less critical actions until you have the airplane trimmed for cruise flight. Don't be fiddling with VOR frequencies or GPS coordinates while you're trying to level onto a course and altitude; those items can wait the extra couple of minutes until you're in cruise.

When you're ready to descend, pick an altitude 1000 feet before your level-off altitude and use that as an ACA "entry point." At 1000 feet above altitude, invoke the sterile cockpit rule. Avoid unnecessary conversation. Delay unnecessary action. Concentrate on what it takes to establish the new level flight attitude, trim the airplane for level flight, and check to make sure critical items are complete before turning to less crucial tasks.

Lastly, once on the instrument approach or within 1000 feet of the ground for landing, once again declare a sterile cockpit, and do only what you need to do to get your passengers, yourself, and your airplane safely on the ground.

Gear extension and verification

If you're flying a retractable-gear airplane, be absolutely certain you extend *and check* the landing gear position at least once before touchdown. It's best to have a pre-planned spot to lower the gear *every time*, such as abeam the landing point in a standard VFR traffic pattern, at the final approach fix of an instrument approach, or when entering the landing altitude-critical area (within 1000 feet of the ground) if flying a non-standard visual approach entry.

It's doubly important not only to extend the gear but also to confirm that it's safely down. Hang on to the gear switch handle until you have a second to check for safe indications. Call "three green lights" out loud as a reminder and verification. On short final, use a GUMPS check, full flap extension, or some other normal action as a reminder to check, and call out "three green lights" a second time, a last-minute confirmation.

Multiple approaches

We've seen how flying an approach a second, third, or even greater number of times can tempt the pilot to "bust" minimums, sometimes with disastrous results. Establish a personal standard operating procedure that you'll only make a second attempt at an approach if (1) you have reason to believe that conditions will improve before the second attempt, making success more likely, or (2) you admit you were sloppy on the first attempt, and you're *sure* you can do better the next time around. Even then, be just as prepared for a missed approach as you should have been the first time.

If (1) or (2) permit you a second attempt and you miss again, go to your alternate or somewhere with better weather. A second try at an instrument approach is a human factors gamble, but a third attempt often proves to be far riskier because it is a symptom of the pilot's desire to land at *that* airport regardless of what it takes. Make "two tries, maximum," part of your personal standard operating procedure.

Cockpit automation

The higher performance an airplane, typically, the more complex its avionics. Take the time to get *very* familiar with cockpit controls and automation. If the airplane has an autopilot, read the supplement, talk to instructors and other pilots familiar with that make and model of autopilot in that make and model of airplane, and practice using the various autopilot modes in clear air, when you're not under stress. To be safe, you need to know how every mode of the autopilot is engaged and functions, and every possible way the autopilot can disable modes or turn completely off, whether commanded by you, by accident, or by system failure.

Similarly, altitude capture equipment, which uses the autopilot to level off at a predetermined altitude without pilot action, is easy to set incorrectly. Get very familiar with the equipment before using it in instrument conditions or busy airspace, and know what abnormal indications look like so you can detect failure before it leads to pitch excursions or an altitude bust.

Regardless, using an autopilot or altitude capture device does not relieve you of your pilot-in-command responsibilities. You must actively monitor the autopilot to make sure it remains engaged and does what you expect it to do. Glance away to run checklists, set up for approaches, scan for traffic, or perform other cockpit chores, but don't delegate flying completely to automation.

Study and practice

Last, checklists and standard operating procedures represent thought-out ways to fly normal and emergency procedures, developed by knowledgeable professionals who were *not* under the pressure of flight when writing the procedures. In effect, checklists and SOPs are like having the airplane designers, engineers, and test pilots along for the ride, giving you their best advice to use in piloting the airplane.

Remember too that regulatory agencies rarely come up with rules and procedures proactively. They don't have the time or, frankly, the imagination to figure out *possible* bad scenarios. Most rules and regulations, then, come from an actual accident experience—they are attempts at keeping us from *repeating* some situation that has *already happened* in the past. In this regard, flying rules are not meant to be broken!

With all this information and expertise available to you, it makes sense to study checklists and procedures regularly and practice unusual or emergency procedures in the cockpit on the ground when you're not under the pressure of actually flying. The cockpit itself is a great "procedures trainer," allowing you to present an emergency and go through the required actions, step-by-step, actually seeing and feeling what you'd do in flight. Envision not just the electrical fire or the engine failure or the radio outage, but continue your practice to its logical outcome, a safe landing. You'll find a simple "alternator failure" scenario, for instance, prompts practice in any number of procedures.

We're almost all taught a "right" way to use checklists: "Read a step, do a step, read another step, do another step." Remember: There are many ways to use a checklist, and whatever way works for you, is safe, and gets the job done is "right." The only wrong ways to use a checklist are those that involve skipping steps—or not to using the written checklist at all.

Once we get comfortable with an airplane or a procedure, it's easy to throw the printed checklist aside and fly strictly from memory. Trouble is, there's a lot going on in most single-pilot airplanes. And we're all subject to the negative effects of workload, health, fatigue, and even our own personality styles. Any of these factors can easily cause us to forget to do something, or to do something wrong.

It's good to try to commit important tasks to memory. However, *always* pull out the checklist as time permits, and confirm you've done everything you thought you did. In a single-pilot environment, there's no one between our own failings and our fate.

Using a checklist is like flying with an expert copilot even when you fly alone. It's easy for age, experience, and ego to lead you astray, but the fact remains that the lists were created by the people who designed and tested the aircraft. Used correctly, the lists remind us to perform tasks we may overlook when outside forces make things complicated in the cockpit. You don't have to be a slave to the "read a step, do a step" method. Just be practical and consistent in your approach.

Where our lives—and the lives of our passengers—are concerned, a little constructive paranoia is a good thing. Use checklists for every phase of flight: start-up, run-up, takeoff, climb, cruise, descent, approach, landing, and shutdown. Checklists are not a crutch. They are a tool that verifies you've covered everything—regardless of interruptions—in a busy, single-pilot cockpit.

Index

About the Author

Thomas P. Turner holds instructor ratings for instruments, and single- and multi-engine airplanes. He has a master's degree in aviation safety, contributes to over 30 aviation magazines, and is a regular speaker at EAA AirVenture and other pilot gatherings. Turner also wrote *Cockpit Resource Management*, now in its Second Edition, and *Weather Patterns and Phenomena: A Pilot's Guide*, now in its Second Edition, both bestsellers in McGraw-Hill's *Practical Flying Series*.